Cognitive Systems Monographs

Volume 13

Editors: Rüdiger Dillmann · Yoshihiko Nakamura · Stefan Schaal · David Vernon

Emanuele Bardone

Seeking Chances

From Biased Rationality to Distributed
Cognition

 Springer

Rüdiger Dillmann, University of Karlsruhe, Faculty of Informatics, Institute of Anthropomatics, Humanoids and Intelligence Systems Laboratories, Kaiserstr. 12, 76131 Karlsruhe, Germany

Yoshihiko Nakamura, Tokyo University Fac. Engineering, Dept. Mechano-Informatics, 7-3-1 Hongo, Bukyo-ku Tokyo, 113-8656, Japan

Stefan Schaal, University of Southern California, Department Computer Science, Computational Learning & Motor Control Lab., Los Angeles, CA 90089-2905, USA

David Vernon, Khalifa University Department of Computer Engineering, PO Box 573, Sharjah, United Arab Emirates

Author

Dr. Emanuele Bardone

University of Pavia
Department of Philosophy
P.zza Botta, 6
27100 Pavia
Italy
E-mail: Bardone@unipv.it
http://philos.unipv.it/emabardo

ISBN 978-3-642-26783-3   ISBN 978-3-642-19633-1 (eBook)

DOI 10.1007/978-3-642-19633-1

Cognitive Systems Monographs   ISSN 1867-4925

Typeset & Cover Design: Scientific Publishing Services Pvt. Ltd., Chennai, India.

Printed in acid-free paper

5 4 3 2 1 0

springer.com

To my brother

Man is the measure of all things.

Protagoras, Greek philosopher (485 BC - 421 BC)

Man is an external sign.

C.S. Peirce, American philosopher (1839-1914)

Preface

In the episode "Homer Defined" from The Simpsons, Homer saves the Springfield nuclear power plant from meltdown. He saves it by performing a children's nursery rhyme that allows him to guess which button should be pressed to avert the disaster. Homer immediately becomes a local hero. Indeed, nobody knows how he managed to prevent the local nuclear power plant from meltdown. He even receives an "Employee of the Month" award from his boss. It is only when he saves a second nuclear power plant by using the same rhyme in front of everybody in Shelbyville that his trick is discovered. So, now everybody knows that it was not omniscient intelligence that saved the entire population, but just a blind guess performed by a dumb man like Homer. The episode ends with Lisa reading the phrase "to pull a Homer" – a new idiomatic expression now entered in the dictionary, meaning "to succeed despite idiocy". This book is about what "to pull a homer" really means to us. More precisely, it addresses the problem of how we as humans succeed despite our *boundedness*.

What the story about Homer tells us is that sometimes we perform well when we think we do not. This is somehow captured by the apparently contradictory statement that sometimes less is more and more is less. As the example of Homer suggests, when we are urged to decide over a certain matter, even a meaningless rhyme like *Eenie, meenie, miney, moe* might be successful or, at least, helpful. Why? Because it makes a problem affordable. Like Homer, we always have a button to press and sometimes we do not know which the right one is. When facing up to unavoidable decisions, anything is always better than nothing.

The example of Homer and his rhyme is quite extreme, but it points to one of the most distinguishing abilities that human beings display, that is the ability of turning almost everything – even a string of meaningless words – into a clue to make a problem affordable in relation to what one knows and, most of all, to what one *does not* know. That is what characterizes humans as *chance seekers*.

Many centuries ago the Chinese military general and strategist Sun Tzu
wrote in his *The Art of War* that when we do not have any idea how to make
the required decision, then "everything looks like important information"
and "it becomes impossible to sort the useless from the useful". This does
not mean that we are completely lost and in the dark. Quite the contrary,
it means that in the absence of *premissory starting points*,[1] ignorance may
turn out to be a chance to be cognitively virtuous. Ignorance should not be
considered simply as absence of knowledge as it might be used as a clue for
making a decision affordable to us. Consider this very simple example. I am
at the airport and I do not know which terminal my flight will take off from.
I could ask a flight assistant to help or I can simply go to one of the terminals
and check the timetable to see if my flight appears one of the monitors. If
not, it means that I am at the wrong terminal. In this case, it is by means of
what remains unknown – my ignorance – that I can make a guess.

There are several other ways in which we can make the best of our ignorance. For instance, there are some things that we think we know simply
because we are *told* that they are so. But actually we do not know that they
are so. For example, when taking advice, we simply trust the person who
is giving it to us, whether we are buying a new laptop or selling our stocks
and shares. Sometimes, we think we know something that we do not, simply
because we are told that it is so by a person who we trust. While on other
occasions we may even watch what other people do and follow the crowd.

It is worth noting here that in all the cases I have mentioned – from homer's
rhyme to taking advice, or not – our ignorance remains *preserved* somehow.
For instance, if a friend of mine says that the new laptop by Apple or Dell is
worth buying because it has certain features, I could follow his advice, and
yet not know anything about either laptop. The same can be said about the
example of the flight. I still do not know my flight terminal, I just know that,
if that were the case, I would know it.

What all these examples share is that they can be easily dismissed or
debunked. That is, it is easy – from an intellectual perspective – to point to
their weakness. In fact, they are all traditionally considered as fallacies. My
ignorance about my flight terminal and the fact that I trust my friend are
scarcely relevant. In sum, they are easy to deploy but also easy to dismiss.
This is basically the rationale of what I call *fallacious* or *biased rationality*.
They are rational insofar as they do not necessarily lead us to a bad outcome,
moreover, they are not dependent on a particular context but, at the same
time, they appear to be quite unreliable.

The main problem of fallacious or biased rationality is that it contrasts
with a human attitude according to which some people are not satisfied with
weak arguments or *it is just so* strategies. They want something better, for
instance, some chances that are well-grounded, more reliable, or at least less

---

[1] I derive this expression from Woods (2009).

weak than others. On some occasions, human beings are able to deploy arguments or strategies that preserve ignorance, yet *mitigate* it.

Ignorance might be a clue, but, as Socrates contended more than two thousand years ago, it can also be a stimulus to *learn* things as we are motivated to formulate arguments and use strategies that are potentially harder to dismiss. My thought is that, when resorting to this kind of "Socratic" ignorance we need to distinguish between what it is *relevant* or *symptomatic* to our decision and what it is not.

This second attitude points to a different form of rationality that takes advantage of the idea of *distributed cognition*. Basically, humans improve their survival strategies by building eco-cognitive structures capable of delivering potentially ever more symptomatic information. It is through various manipulations of the environment that we gain new and more reliable chances which can be used to *de-bias* our rationality. Indeed, a children's nursery rhyme or a trustworthy friend allow us to make decisions affordable, even when they would not otherwise be so. At the same time however, through the laborious activity of cognitive niche construction, we come up with situations in which we are *better afforded* by our environment, and thus biases or fallacies cease to be appealing. In this sense, our environment is a source of selection pressures over human life, but also a storehouse of cognitive chances, namely, affordances, which are potentially more relevant to our survival and prosperity.

Manipulation of the environment – a hypothetical activity – unearths affordances that, once stored in our various cognitive niches, can be accumulated and contribute to de-biasing our rationality. Indeed, this is not a permanent result. There is no method of securing successful affordances to our genome. Our cognitive niches – and all the extensions of our rationality – may be enriched, but they can also perish or collapse. In this sense, our rationality is still bounded. That is, the activity of distributing our cognition does not lead to a complete de-bounding of human cognitive system. However, it contributes to move the bounds of cognition.

The Structure of the Book

One of the facts that I take for granted in this book is that human cognition is bounded. Human cognition is bounded when it falls short of omniscience. In my view, this simple statement warns us to adopt a cognitive agenda, which does not seriously take into account what people actually do, before considering what they *should* do. And what do people often do? They make mistakes, lots of mistakes. Mistakes are basically symptoms of what it is going on within our cognitive system. In the first chapter *Fallacies and Cognition: The Rationale of Being Fallacious* I will develop this idea presenting the case of fallacious reasoning as an example of the importance of accounting for mistakes. The frequency and ubiquitousness of fallacies – traditionally considered as patterns of poor reasoning – is symptomatic of the fact that human cognition has a strong commitment to *cognitive economy*. When operating in cognitive economy, what appears to be clearly erroneous may turn out not to be so. Fallacious rationality has its appeal and investigating what it precisely consists of is an imperative task. The main thesis I will develop is that the importance of fallacies is *symptomatic* in the sense that they are symptoms of the way cognition works.

In the second chapter *Bounded Rationality as Biased Rationality: Virtues, Vices, and Assumptions* I will look carefully into the kind of boundedness we are limited by. In doing so, I will integrate the issue of fallacious reasoning into a wider debate concerning biased rationality. In that chapter I will show the vices and virtues of biased rationality. I will discuss when biases and fallacies are good, but also when they uncover unreliable or at least "maladaptive" solutions to our survival and prosperity. Some questions will then be explored in more detail. What does being cognitively bounded really mean? As Woods arguably noted, "we survive, we prosper, and from time to time we build great civilizations". How could that be possible given our limitations? How could we possibly account for the amazingly successful outcomes humans sometimes bring about?

Our commitment to cognitive economy does not imply that the cognitive assets humans have cannot be improved and extended beyond previous limitations. This is basically the idea that while we cannot get rid of our boundedness it is not to say that we cannot reach something better. For the problem is not that of the bounds (that we do have) but that of *their instability*. Bounds are not fixed once and for all but are in constant movement. For example, they move (a) in relation to representations of the problem and of alternatives, (b) in respect to resources they use (paper, pencil, computer, figures, tables, books, reports, etc.), (c) because of the creative activity of our brain (here heuristics operate, for example), (d) as emotional states intervene, (e) together with moral values, and so on. To view the bounds of human cognition as moving is made possible by the assumption that our cognitive system is distributed. This will be discussed in the third chapter *Moving the Bounds. Distributing Cognition through Cognitive Niche Construction*. The main thesis that will be illustrated here is that human beings do not actually hold a complete representation of their environment. Conversely, they use the environment itself as a representation by manipulating and even creating it so as to find room for new *cognitive chances* which were not immediately available. This idea of human cognition as a chance-seeking system will be developed within an evolutionary framework based on the notion of cognitive niche construction. According to this theory, the high level of plasticity exhibited by humans is explained by the fact that humans are powerful *eco-cognitive engineers*. The accumulation and, most of all, the persistence of modifications on the environment is what grants humans an additional source of information, not delivered through genetic material, that is in fact fundamental for behavior control.

It appears to be a circular argument to claim humans turn environmental constraints into ecological chances when facing the challenges posed by the environment itself. That is not the case, as we assume that organisms (including humans) adapt to their environment, *and vice-versa*. This will be illustrated in the fourth chapter *Building Cognitive Niches: The Role of Affordances*. What I will argue is that human cognitive behavior consists in *acting upon* those anchors to which we have secured a cognitive function, via cognitive niche construction. Those anchors are basically *affordances*. Affordances are a way of measuring or representing the environment with respect to the action capabilities of an individual. Here, again, humans do not hold a complete internal representation of the environment but they use the environment itself as a model insofar as they can immediately access it in terms of those action capabilities, which emerge in the interplay between humans and their environment. The notion of abduction will contribute to making our point bolder. Going beyond a sentential and computational dimension of abduction towards including it in a broader semiotic one, I will argue that affordances can be related to the variable (degree of) *abductivity* of a configuration of signs. This will be of help when illustrating the evolutionary dimension of affordance detection and creation. I will argue that humans have at their disposal a *standard* or *pre-wired* endowment of affordances, but at the same time they can extend and modify the range of what can offer them affordance through the development of appropriate cognitive abductive skills.

As already argued, what is crucial for making plasticity work is to turn environmental constraints into *ecological chances*. This transformative activity is at the core of our proposal, which is to view human cognition as a chance-seeking system. We build and manipulate cognitive niches so as to unearth additional resources for behavior control. This activity of eco-cognitive engineering is basically what best describes our idea of *learning* as an *ecological task*. In the fifth chapter *The Notion of Docility: The Social Dimension of Distributing Cognition* I will present the notion of docility. First introduced by Herbert Simon, I develop the original notion arguing that docility is that kind of disposition underlying those activities of ecological learning. As most of the resources we benefit from are stored externally, docility is supposed to facilitate the delegation and exploitation of cognitive chances secured to cognitive niches. From an evolutionary perspective, docility is an adaptive response to the increasing cognitive demand on those information-gaining ontogenetic processes like learning, resulting from an intensive activity of cognitive niche construction. In this sense, docility makes people more inclined to overcome their ignorance by means of learning. It facilitates information sharing and accumulation.

The last chapter *Seeking Chances. The Moral Side* is a sort of appendix. It is an attempt to integrate morality into the distributed framework presented in the previous chapters. This is not an attempt to articulate a complete moral theory on the basis of our proposal. Rather, it is meant to offer its practical application to moral reasoning. What I will discuss is that the mechanism underlying chance-seeking activities may capture some important features of moral reasoning. Basically, I will present the thesis, first introduced by Lorenzo Magnani, according to which morality is a distributed phenomenon. Morality is distributed in the sense that even our moral agency is continuously shaped and reshaped by the activity of niche construction. Various technological artifacts, but also institutions and language itself, extend our capacity to discern moral values and cope with situations which would require a moral commitment. I am far from developing a unified approach which would combine cognition and morality, but we believe that this would be a valuable starting point.

I started to think about the research covered in this book while studying as a PhD student in Philosophy at the University of Pavia, Italy. Preparation of this work would not have been possible without the resources and facilities of the Computational Philosophy Laboratory (Department of Philosophy, University of Pavia, Italy). This project was conceived as a whole, but as it developed various parts became articles, which have now been revised and integrated into the current text. I am grateful to Springer for permission to include portions of previously published articles. Parts of Chapter 1 were previously included in: E. Bardone and L. Magnani, "The Appeal of Gossiping Fallacies and its Eco-logical Roots" in *Pragmatics and Cognition*, 18(2), 2010.

For valuable comments and discussion of a previous draft I am particularly grateful, first of all, to Lorenzo Magnani (University of Pavia, Italy) who has been an ideal supervisor over the years I have spent as one of his PhD students. He is a great mind, a great human being: it will be very hard to repay his kindness. A very special thanks goes out to Ester Võsu (University of Tartu, Estonia) who has really extended my cognition during the last stages of preparation of this volume. My work has benefited from discussions with several people over the last four years. Chiefly, Davide Secchi (University of Wisconsin, La Crosse, US), Tommaso Bertolotti (Universty of Pavia, Italy), and Bernardo Pino Rojas (University of Santiago, Chile) who all deserve a special mention for having challenged my thoughts and ideas, which otherwise might have become a graveyard of circular arguments. My thanks and appreciation goes also to all those students who have taken my course in Cognitive Philosophy at the University of Pavia and road-tested the ideas in this book. I would thank Claudio Pizzi (University of Siena, Italy) and Liliana Albertazzi (University of Trento, Italy) for providing me with valuable comments and remarks on a previous draft of the book. Finally, a special thanks to and Kai Pata (Tallin University, Estonia), Roberto Feltrero (UNED, Madrid, Spain), Merja Bauters (University of Helsinki, Finland), Kristian Bankov (University of Sofia, Bulgaria).

Some sections have been written in collaboration with Lorenzo Magnani and Davide Secchi: sections 1.1, 1.2, 3.1, 3.3, 4.1, 4.2, 4.3 with Lorenzo Magnani, sections 2.1 and 2.2 with Davide Secchi.

In conclusion, I recognize that this research would not have been possible without the financial assistance of the Italian Ministry of Education and the project "Dottori di ricerca e mondo del lavoro" (2009), University of Pavia, Italy.

November 2010

Pavia

Contents

Chapter 1
Fallacies and Cognition: The Rationale of Being Fallacious

Introduction

The opening chapter aims to set the scene, presenting some of the main issues to be discussed in the following chapters. Basically, this chapter is about human beings as *error-prone* creatures. It is a matter of fact that people make mistakes, many mistakes. No matter how trivial this consideration may appear, most researchers fail to show sufficient interest in people's mistakes.[1] In decision-making theory, for instance, mistakes are considered as merely accidental. That is, researchers do not incorporate this kind of error into their accounts or theories. Indeed, we build up theories to avoid faults. But sometimes the fact we make mistakes and the way we make them, may be symptomatic of the mechanism that lies behind decision-making or problem-solving activities.

Indeed, the term "error" is very broad in its semantics, since it encompasses for instance, perceptual errors, mechanical errors, faulty memories and factual misinformation. In this chapter, I shall consider a narrower class of errors traditionally called *fallacies*. Logic has always been committed to helping people make sound judgments, and fallacy appears to be a failure with respect to this target. Also, if fallacies may warn us when logic fails to meet cognition, as far as I am concerned here, this failure turns out to be symptomatic of the way human cognition manages various resources (logic included) to make decisions and solve problems.

The thesis I will start off with is that a fallacy is a form of reasoning that for some applications is *bad* and for others is *good*; roughly speaking, it is bad for logic but, for instance, good for surviving. So, in the first part of this chapter, I will

---

[1] An important exception to this is constituted by the work pioneered by Tversky and Kahneman [1983; 2003] on the psychology of judgment, reasoning and decision making. They explicitly focused on the extensive use by humans of heuristics and short-cuts, and how their judgment is drastically influenced by biases, leading them to deal with decision making under uncertainty. Bias research has been become influent especially in the study of probabilistic and deductive reasoning. For a recent review on the matter see Evans [2002]. On the relationship between fallacious arguments and the work of Tversky and Kahneman, see for instance Hintikka [2004].

E. Bardone: Seeking Chances, COSMOS 13, pp. 1–19, 2011.
springerlink.com © Springer-Verlag Berlin Heidelberg 2011

discuss some of the limits of traditional logic in approaching the problem of fallacy. In doing so I will take advantage of the theoretical framework developed by Gabbay and Woods in their recent publications.

In the section 1.2, I will discuss specific classes of fallacies, namely, the *argumentum ad hominem*, the *argumentum ad verecundiam*, and the *argumentum ad populum*. I will argue that they share a common *taste for gossip*. That is, I will maintain that these fallacies prove extremely successful because they make use of beliefs and/or knowledge about other people.

1.1   The Appeal of *Being-Fallacious*

From the perspective of classical logic, a fallacy is a pattern of poor reasoning which appears to be a pattern of good reasoning [Hansen, 2002]. Two disciplines are involved in fallacious reasoning: formal logic, which is mainly addressed to 'formal fallacies", and informal logic, that describes the so-called 'informal fallacies". First of all, we can say that the validity of a deductive argument depends on its form; consequently, formal fallacies are arguments which have an invalid form and are not truth preserving (for example the fallacies of affirming the consequent and of denying the antecedent). On the other hand, informal fallacies can be other modes of reasoning whose failure is not strictly based on the type of argument (for example the "ad hominem argument" or the "hasty generalization"). Even if there is no agreement upon an established taxonomy, the fallacies discussed in the context of informal logic typically still include formal fallacies (which are of course also discussed in formal logic) such as the famous fallacy of affirming the consequent and the fallacy of denying the antecedent, but also the proper informal fallacies such as "ad hominem" (against the person), "slippery slope", "ad baculum" (appeal to force), "ad misericordiam" (appeal to pity), and "two wrongs make a right".

Even if the conception of good inference is usually able to model many kinds of real human argumentation, its appeal to true premises is ill suited to many contexts which are characterized by the presence of hypothetical and uncertain beliefs, by strong disagreement about what is true and false, by ethical and aesthetic claims which are not easily categorized as true or false and, finally, by variable contexts in which dramatically different assumptions may be accepted and rejected [Magnani, 2009].

Let me consider the case of the so-called "hasty generalization".[2] As I will demonstrate a hasty generalization is an example of fallacy; a fallacy that can however lead the cognitive agent – in spite of its fallacious character – to fruitful outcomes, thus not seeming to be a fallacy at all. As Woods put it "hasty generalization is sometimes a prudent strategy especially when the risks are high, even though the haste of the generalization might attract a charge of fallaciousness" [Woods, 2004, p. 13].

---

[2] For a detailed treatment of hasty generalization, see for instance Walton [1999b].

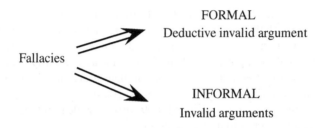

FORMAL
Deductive invalid argument

Fallacies

INFORMAL
Invalid arguments

Hasty generalization occurs when a person (but there is evidence of it in animal cognition too, for example in mice) infers a conclusion about a group of cases based on a sample that is not large enough. It has the following form:

- Sample S, which is small, is taken from the group of persons P.
- Conclusion C is drawn about the group P based on S.

It could take also the form of:

- The person X performs the action A and has a result B.
- Therefore all the actions A will have a result B.

The fallacy is committed when not enough A's are observed to warrant the conclusion. If enough A's are observed then the reasoning is not fallacious, at least from the *informal* point of view. Males, when driving their cars, have probably quarreled with a woman driving her car and, while quarreling, they have argued (or shouted) "all woman are bad drivers!". This is our case of fallacious reasoning.

Thus small samples are likely to be unrepresentative. Another simple case is the following. If we ask one person that recently met a lot of Italians what he thinks about the new Italian electoral system, his answer clearly would not be based on an adequately sized sample to determine what Italians in general think about the issue. This is because the answer given is based only on a reduced number of experiences and such judgment is not sufficient to provide a generalization about the matter in question. It is not good to generalize from only one available sample.

People often commit hasty generalizations because of bias or prejudice. For example, someone who is a sexist might conclude that all women are unfit to fly jet fighters (or to drive a car) just because one woman crashed, in either case. People also commonly commit hasty generalizations because of laziness or sloppiness. It is very easy to simply jump to a conclusion and much harder to gather an adequate sample and draw a justified conclusion. Thus, avoiding this fallacy requires minimizing the influence of bias and taking care to select a sample that is both large enough and meaningful.

Moreover, I can recognize another important occurrences. I have said that people commit errors and are hasty generalizers because of prejudice, mindlessness, bias, and so on. What I am trying to argue is that a hasty generalization is not always a bad generalization for two reasons. The first is that, for obtaining true conclusions, hasty generalizations might be good if the *result* of the generalization we made coincides with the result of a good generalization in the philosophical sense of induction, or in the sense of inductive logics. We call this case the "causal" truth-preserving feature of hasty generalization, meaning that the truth stated in the conclusion can be preserved, even though that is not a permanent feature of a hasty generalization. The second reason is that, in some sense, even if we do not reach good conclusions, not exploiting the casual truth preserving feature, we can say that hasty generalization is good in some sense although obviously not in that of classical logic. I will now try to understand what it can be achieved [Woods, 2004].

From an intuitive perspective, Woods [2009] offered a clear-cut view on fallacy describing it with respect to four main features:

1. A fallacy is a pattern of poor reasoning. That means it is *erroneous*. It leads us to a conclusion by means of poor reasoning.
2. A fallacy is a pattern of poor reasoning that, however, looks good. That is, it is *attractive* and *seductive*, since it can be easily smuggled as a good argument sometimes.
3. A fallacy is also *universal* in its occurrence. That is, everybody is got used to commit fallacies, no matter where she lives or what culture she belongs to.
4. A fallacy is *incorrigible* in the sense that people display high level of post-diagnostic recidivism; committing fallacies is a incorrigible disposition or mind-set.

Erroneous, *attractive*, *universal*, and *incorrigible* are all adjectives that clearly depict an intuitive notion of fallacy, labeled by its proponents with the acronym *EAUI*.

1.1.1 The Agent-Based Perspective

The idea Gabbay and Woods [2005] have put forward is that logic should look at what is done by a cognitive agent. That is, it should pay more attention to the cognitive agent's concrete ways of reasoning. The fact that humans are quite prone to committing fallacies cannot be disregarded, if the aspiration of logic is that of being an aid to human cognition. This kind of approach is what Gabbay and Woods [2001] labeled as *agent-based* logic.

Agent-based logic consists in describing and analyzing the reasoning occurring in problem solving situations by looking to what a practical agent actually does, and not what she/he should do or is supposed to do. We arrive to what has been called the "Actually Happens Rule" [Woods, 2005, p. 734] that states "to see what agents should do we should have to look first to what they actually do and then, if there is particular reason to do so, we would have to repair the account". This particular approach assumes that the description of reasoners comes before the description of reasoning. In doing so, we indeed introduce an element of psychologism in logic [Woods, 2009]. Now the question is: what is "beings-like-us" meant to be? The agent-based perspective on logic aims at giving a principled description of certain aspects of a cognitive system, which is depicted as a triple of 1) an agent C, 2) cognitive resources R, and 3) a cognitive target J. Having this in mind, practical agents seek to attain the targets they set with the resources they can reasonably get and/or have at hand. Therefore, practical agents operate in *cognitive economies*

[Woods, 2009], where the agent access to cognitive resources encounters limitations such as:

- bounded information
- lack of time
- limited computational capacity.

Hence, the "beings-like-us" that Woods [2005] describes in his "Epistemic bubbles" discharge their cognitive agendas under press of incomplete information, lack of time, and limited computational capacity. We can consequently say that cognitive performances depend on information, time, and computational capacity. An *agent-based logic*, as a discipline that furnishes ideal descriptions of *agent-based reasoning*, returns to be thought of as a science of reasoning and considered *agent-centered, task-oriented, and resource-bound*.

1.1.2 Proportionality and Relativity of Errors

As already mentioned, fallacies are bad from certain perspectives but good from others. In this sense, an argument turns out to be erroneous in relation to a standard or a set of standards it fails to meet. For instance, induction might be good if the result of the generalization we made coincides with the result of a good generalization in the philosophical – for example Millian – sense of induction (or in the sense of inductive logics). In this case Millian methods for induction is the standard set for describing good arguments [Magnani, 2009]. Now, think of a toddler that touches a stove in the kitchen for the first time [Woods, 2004]. His finger is now burned because the stove is hot. From this evidence, the hastily generalizing toddler thinks that all stoves are hot and decides not to touch stoves anymore. In this case, the toddler's pattern of reasoning, namely, a hasty generalization, is extremely poor, if we adopt the Millian methods for induction as our standard. However, if we adopt a much less cognitively pretentious standard (for instance, the one related to "do not burn your hand"), his hasty generalization does not appear fallacious, but it is a good example of pragmatic reasoning.

It is noteworthy that the setting of a standard depends on, and is proportionate to, the target (or interests) and the resources (information, computational capabilities, etc.) the practical agent has at his disposal. This last consideration depicts a conception of fallacy which pays more attention to the relativity of error and its proportionality to the resources available at a given time. That is, committing a mistake is much closer to the selection of cognitive protocols that exceed an agent's competence than the mere violation of a given standard which is impossible to meet just like in the case of the toddler's hasty generalization. As Woods [2009] put it, "something is an error of reasoning only in relation to the reasoner's cognitive targets and the attainment-standards that they embed".

In order to better clarify this point, the distinction between the *individual* agent and the *institutional* agent is useful. As Woods put it [2009]:

[...] hasty generalization is not a fallacy when committed by human individuals, but it might well be a fallacy when committed by institutional agents such as Nato, NASA or MI5, or cultural agents such as Soviet physics in the 1960s or Silicon Valley in the 1980s. This turns out to be a vital distinction for our case, for it is a distinction driven by the fact that agency types – whether individual, institutional or cultural – are largely defined by the cognitive assets on which they are able to draw in the discharge of their reasoning tasks.

The toddler is, indeed, an individual agent. Conversely however, if we think of science, for instance, it is an institutional agent. According to what has already been pointed out, the hasty generalization made by a toddler after touching a burning stove is not an error; given the task of not being burnt for a second time, the hasty generalization is a kind of reasoning that is fruitful because, being a prudent strategy, it embeds the canons of *strategic rationality* in the sense of the "strife for survival". But what if science as a cultural institution made the same inference? That would be a terrible mistake, indeed.

Compared with the toddler, science as an institution can exploit many more resources; in addition, the targets set by science embed much higher standards than the toddler's. To science, it is not sufficient to state that "every time the toddler touches the stove, it will burn". Science works another way: it has to analyze in detail the causal chain that explains the conditions under which the toddler's hand will be burnt by the stove, for example. The apparatus science employs far exceeds that exhibited by the toddler, whose "scientific apparatus" is limited to his hands or, at least, to his ten fingers (that means he would have a maximum of ten tries). For example, a simple temperature sensor used by scientists in a laboratory gets better feedback than the toddler can get. Science can also measure the interval of time after which the stove gets cold, and therefore can be touched safely and without any harm.[3]

The idea I will develop in the following is that the cognitive appeal of fallacious arguments is rooted in human evolution belonging to what Woods [2004, p. 8] called the *rational survival kit*. This *rational survival kit* can be considered as the result of evolution, in which basic equipment, genetically and culturally endowed with certain abilities, has been selected and learnt for survival and reproduction. That is, fallacy appears to be a kind of error which is committed "when the agent's cognitive devices are functioning as they should" [Woods, 2009]. Therefore, the importance of the fallacies is *symptomatic* in the sense that they are symptoms of that rational survival kit, which has been forged and shaped by evolution, and of course by its

---

[3] It is worth to note that this kind of reasoning is here ideally depicted and reconstructed as a case of hasty generalization, in the framework of a sentential inferential framework. An objection could be provided by the psychologically realistic observation that an actual child does not really perform a propositional inference as become averse to hot stoves, but instead he is basically directed by some emotional abductive reactions. However, this does not impede us to choose the other perspective of "reconstructing" in a non psychologically way but in an inferential/propositional scheme the hypothetical inference performed, so revealing its fallacious character. For a complete treatment on the matter, see Magnani [2009].

limits. The fallacious nature of fallacies can also be termed *derivative* [Woods, 2004, p. 9], since fallacies do not derive their fallacious nature from themselves, but from the human survival kit.[4]

Now, if fallacies are symptomatic of our survival kit, the question arises, what are those skills that make this kit appealing? What do fallacies tell us about those skills? These questions concern a wider scope of issues than the one I wish to deal with here. My aim is more modest. What I want to do is to consider a narrower number of fallacies which are symptomatic of ways of reasoning that are prominently *social*. This will allow us to introduce the eco-cognitive dimension involved in fallacious reasoning which, in my view, is what makes them appealing. In the following section, I provide a description of three fallacies that we think can be useful for this purpose.

1.2 How to Make Use of Social Characters

It is hard to arrive at a definite and complete list of typical fallacies, therefore I conventionally assume the list provided by Woods [2004]. Woods listed eighteen items – the so-called *Gang of Eighteen*. More precisely, I refer to three main types of fallacies:[5] *argumentum ad hominem* (or *argument against person*), *argumentum ad verecundiam* (or *appeal to authority*), *argumentum ad populum* (or *appeal to popularity or bandwagon*). All these three fallacies are traditionally considered as examples of a broader category called *ignoratio elenchi*.[6] The three fallacies employ a general pattern of reasoning based on the introduction of some irrelevance that does not deal with the matter in discussion.

1.2.1 Argumentum ad Verecundiam

The so-called *argumentum ad verecundiam* is based on the appeal to an authority acknowledged as such in order to support or boost a certain position rather than another. Consider the following example. Andrew Keen [2007] wrote a book, *The Cult of Amateur*, in which he violently attacks the culture that the Internet and the Web are nurturing. He argues that new technologies, such as blogs, social networking sites like MySpace, self-broadcasting tools like YouTube, etc., are glorifying

---

[4] As it will be clearer in Chapter 3 the evolution of this particular survival kit is better described as resulting from the co-evolution of biological and cultural systems. In fact, the term evolution is used here in a loosely sense, as the evolution of the survival kit we are talking about cannot be entirely represented as a Darwinian system.

[5] What I am going to present here is not meant to be a complete treatment of these three arguments. Rather, it is just a brief sketch, which will be of that importance in order to make my point concerning the relationship between some fallacious arguments and gossip. For a more exhaustive description of the *argumentum ad verecundiam*, see Walton [1997] and Goodwin [1998]; of the *argumentum ad hominem*, see for instance Walton [19998] Metcalf [2005]; of the *argumentum ad populum*, see for instance Walton [1980].

[6] For a detailed account of the role of irrelevance in argumentation, see Walton [2004].

and celebrating the so-called *cult of the amateur*: that is, through breaking up all the traditional intermediate layers between the editor and the users, these new technologies are encouraging everybody to become a source of information and entertainment about a certain issue or topic, no matter if he is an expert on the field he is writing about, or not. Conversely, he claims a well informed public should not rely on amateurs, but talented intermediaries, like professional editors and journalists, for example. In putting forward his thesis against amateurism, he made an example, which concerns an eighty-minute movie called *Loose Change* on 9/11 conspiracy theories which sprang up on the Net some years ago:

> The "claims" made by *Loose Change* were completely discredited in the final report of the 9/11 Commission, a report that took two years to compile, cost $15 million, and was written by two governors, four congressmen, three former White House officials, and the two special counsels. [Keen, 2007, p. 69]

In this case, the claim made by Keen is a fallacy based on the *appeal to experts*. It is fallacious because he does not reject the theory presented in the movie *Loose Change* by referring to evidence and incoherencies; conversely, he simply posits that the conspiracy theory presented in the movie is false, because the *9/11 Commission* has reported quite the contrary. To boost this position, he simply lists some details related to the report, for example, the cost of the commission, its composition and duration. Of course, all these details are irrelevant to assess what really happened that day. Generally speaking, the *argumentum ad verecundiam* follows a general pattern of reasoning that can be described as follows:

- A given person *P* thinks *b* is true;
- *P* is supposed to be an expert on the area which *b* belong to;
- Then, we have a good reason to think *b* is true

Under conditions of limited information and computational capabilities, all the irrelevant details cited by Keen about the *9/11 Commission* may contribute to better direct our opinion. That is, he is assuming that a commission report is a reliable and official source of information, which should be more trusted than any other source.

More generally, the fact that people lean on persons who are publicly acknowledged and esteemed as experts in their respective fields is fundamental to facilitate a wide range of activity. Consider, for example, learning: learning is a process that would not be possible without a certain relation between the instructor and the disciple. Since we cannot know everything and acquire all the information required to make a sound judgment, we usually take comments or suggestions that come from people we esteem seriously; and that is the appeal to experts.

In science the *argumentum ad verecundiam* is also often deployed. Consider, for instance, the so-called "Matthew effect", named after the *Gospel of Matthew*: humans tend "to give credit to already famous people". This particular effect has been empirically tested, since Robert Merton [1965; 1996] popularized it. Of particular interest is a well-known study conducted by Lewontin and Hubby [1985] in which they reported a curious case of *ad verecundiam* in science.[7] Lewontin and Hubby analyzed the citations they got back from two papers they co-authored. The two papers were both on the same matter, with the order of authors alternated, and both qualified as "citation classics". What they reported is that the most cited between the two papers was the one in which the first author was Lewontin, who was also the most well known in the scientific community at that moment.

It is worth noting that the *argumentum ad verecundiam* concerns other people, but also inanimate objects like those we use in everyday life. For instance, we usually take for granted the first results we get back from a web search using tools like "Google" or "Yahoo!". When visiting a new place we make use of external aids like a route planner, GPS, maps, and/or directions. Sometimes, we prefer to follow the signs and directions we encounter on the road rather than relying on our memory that can be faulty.

1.2.2 Argumentum ad Hominem

Consider another fallacy called *argumentum ad hominem*, in plain English, *argument against a person*. Here is an example. Some months ago, Britain's Channel 4 aired a documentary on global warming entitled *The Great Global Warming Swindle*. The content of the documentary immediately raised a number of doubts about its scientific reliability. Opposing the thesis put forward in the documentary, Bill Butler made a detailed investigation about the experts, who were interviewed in the movie and who posited that global warming is a fraud based on bad science. This is what he wrote:

> The pseudo-documentary implies that the other people who appeared are knowledgeable experts in their fields. In practice, their best expertise seems to be wrangling payments from large energy companies – especially anti-environmental organizations.

Then, he went on listing the *curricula vitae* of those "alleged" experts, who appeared in the movie. He wrote:

> [...] Despite the caption on the programme, Singer has retired from the University of Virginia and has not had a single article accepted for any peer-reviewed scientific journal for 20 years. His main work has been as a hired gun for business interests to undermine scientific research on environmental and health matters [...].[8]

Indeed, this kind of argumentation is fallacious, since it introduces some irrelevant information in discussing the issue of global warming. Whether or not an expert is on the payroll of an energy company cannot constitute a *probative* basis of evidence for arguing in favor of or against global warming [Walton, 2004]. As a matter of fact, other proof does not matter here, for instance, the CO_2 concentration in the atmosphere, the solar activity, the historical global temperature data, and so on.

---

[7] For a recent discussion of the Matthew Effect in science, see Strevens [2006], who discusses at length both normative and descriptive issues related to that effect.

[8] http://projectearthnews.blogspot.com/2007/03/great-channel-four-swindle.html

Depicting a person putting forward a certain view as a pseudo-scientist or, even worse, as a fraudulent person does not prove him wrong. Generally speaking, the *argumentum ad hominem* follows this general pattern:

- A given person *P* holds belief *b*;
- unfavorable information about *P* is presented;
- Then, *b* is not acceptable.

However, although a personal attack proves nothing about the matter in discussion, from the perspective of a practical agent the introduction of irrelevant information may turn out to be a valuable resource. The *argumentum ad verecundiam* bases its appeal on the release of good information about the person we are listening to, whereas the *argumentum ad hominem* serves the opposite purpose: to unearth that information from which we can infer that the person is not trustworthy. In the example I cited above, it is clear that a person, who is on the payroll of an energy company, may be more inclined to deny the negative consequences on the environment brought about by the kind of companies he actually works for. Of course, this is a prudent strategy that is selected under conditions of limited information. Since we cannot go over all the details of global warming theory, we have to depend on the say-so of others; therefore, we are more inclined to support the position held by those who are – or appear to be – more trustworthy.

1.2.3 Argumentum ad Populum

People tend to believe what the majority believe. When they appeal to this belief, they are using an *argumentum ad populum*. Consider the following case. Over recent years, climate change has attracted a great deal of attention from various political organizations. The relation between scientists and policy-makers has become more and more salient, given the threat that global warming is supposed to hold for the entire human race. Now suppose that a person would claim:

> There are no scientific publications denying global warming. Therefore, I think global warming is really happening and we should act to prevent future damages.

This is an example of the *argumentum ad populum*. This fallacy is based on the appeal to what the majority of people say: in this case, the argument is merely based on the appeal to the consensus among scientists. Roughly speaking, what the person posits is that global warming is likely to be true and deserves public concern, since the entire scientific community agrees that our planet is getting ever warmer. This argument can be resumed as follows:

- The majority of people *M* think *b* is true;
- then, we have a good reason to think *b* is true.

This kind of argument is fallacious for two main reasons. First of all, even scientists might go wrong, because science as a human enterprise is fallible. Secondly, the fact that the entire scientific community agreed upon a given theory is a statement that has nothing to do with global warming; is consensus something capable of inducing climate variations the way CO_2 emissions or solar activity do? Not at all: there is no causal chain between the scientific consensus and the rise of CO_2 concentration in the atmosphere, for example. From a scientific point of view, it is simply irrelevant.

However, it is noteworthy that the appeal to the consensus among scientists is still attractive and seductive. For instance, it is assumed that scientists follow the scientific method, which has been proven over more than five centuries to be one of the most reliable institutions for producing and transmitting knowledge. Science is still fallible and scientists might be wrong, but the consensus reached among scientists is certainly more trustworthy than the one reached in a pub or even in parliament, in which private interests can easily enter the scene. Science exhibits a wide range of public procedures that should lessen the negative impact of non scientific concerns; the articles published in the best ranked international scientific journals are peer reviewed, experiment results cannot be kept secret, but must be released publicly, etc.

The issue of trustworthiness is therefore crucial for beings-like-us, who cannot look into all the scientific reports or results that are monthly or even daily published in specialized journals or magazines. In the case of the *argumentum ad verecundiam*, the appeal to experts is the source of our tendency to trust the *say-so* of others; in the the case of the *argument ad hominem*, what counts is unfavorable information about the person who actually holds that view; finally, in the case of the *argumentum ad populum*, we are naturally inclined to trust what the majority of people thinks, because we are ultra-social creatures [Richerson and Boyd, 1998; Richerson and Boyd, 2005]. That is, human cognition has been biased towards social problem-solving throughout all evolution [Sperber and Mercier, 2010; Mesoudi *et al.*, 2006; Adolphs, 2006; Humphrey, 1976; Whiten and Byrne, 1997].

1.2.4 The Question of Irrelevancy and Fallacy Evaluation

As examples of *red herring*, all the fallacies I briefly surveyed introduce irrelevant information about a certain issue of debate. Although they do not provide conclusive evidence to support or reject a certain claim the way scientific statements do, they tell us something interesting about how humans build up their arguments and reasons: people committing these fallacies rely on information about other human beings in their reasoning. That is, they do not follow certain logical procedures that eventually lead them to correct conclusions. But they simply make use of others as social characters. For example, being an authority, being an expert, being part of a class, etc., becomes the substitute for more direct evidence to support a certain claim or make an argument more appealing. Although all these arguments remain fallacious in their nature, they are somehow good. Indeed, they are not good from a logical perspective, because they are bad arguments, but they can be successful from the agent-based perspective we illustrated in section 1.1.1. That is, they may provide good solutions given the various constraints (bounded information, lack of time, limited computational capacities) humans are subjected to.

Interestingly, Tindale [2007] argued that the evaluation process of a red herring is an important step towards its identification, because upon occasion the arguer does not usually jump to another issue bearing no relation to the one at stake. As already noted, people are used to introducing irrelevant information with respect to the issue under discussion. But when may we say that information introduced is relevant? This seems to suggest that the fallacious nature of a red herring rests on the possibility of proving that a *real* shift in issue has occurred. It is noteworthy that in any communicative context people continuously *strive to be relevant* [Dessalles, 2000; Tindale, 2007; Grice, 1975]. In all the *"ad* arguments" we have briefly illustrated the shift in an argument is symptomatic of the ability to make use of various resources people may have at their disposal. That is, introducing some irrelevancies is not only a fraudulent attempt to divert audience attention, but also the attempt to look for new and potentially relevant connections.

The fact that an argument makes use of irrelevant information is commonly considered as fallacious. However, this statement may turn out to be problematic, as it is not always obvious when and why a piece of information is irrelevant and what makes it as such. Magnani [2009, p. 411-414] argues that both argument evaluation and argument selection are often the fruits of an abductive appraisal; that is, striving to be relevant is guided by abductive skills, which both the arguer and audience employ to select convincing arguments on the one hand, and to evaluate the competitive narratives at stake, on the other.

Abduction is defined by Magnani [2001] as the process in which a hypothesis is created/selected and then evaluated.[9] As far as I am concerned, the notion of abduction may shed light on the mechanism behind argument evaluation and selection, and thus develop the approach put forward by Tindale [2007]. The abductive dimension of argument evaluation can be described as follows. As regards the arguer, he aims to put forward that argument, with which he might best convince his audience that he is right. Here, the abductive skills involved are those required to make the proper guess regarding those clues or the information that his audience will make use of.

For instance, if one is delivering a speech on the relationship between science and religion to a group of Catholic students, one might guess that an appeal to the Pope as an authority would be more likely accepted than an appeal to an atheist. In contrast, an appeal to the philosopher of science Paul Thagard would be more likely to be accepted than any other when discussing the explanatory dimension of scientific reasoning, given that Thagard is a well-known expert in that field. Generally speaking, sophisticated abductive skills can lead the arguer to generate or select the most convincing argument with respect to the nature of the given audience.

As for the audience, the abductive skills are those employed to evaluate whether a certain argument is coherent. Let me return to the example of Paul Thagard. Suppose that the audience attending the discussion consists of mainly PhD students, interested in philosophy of science, and suppose that the topic for discussion is the role of explanation in science: in such a case, it is likely that the students would

---

[9] Abduction will be treated in detail in Chapter 4.

evaluate the appeal to Paul Thagard as a good argumentative move. They know, for example, that Thagard has done much of his research precisely on the role of explanation in scientific reasoning, he has published widely on the topic, he has been frequently quoted and he has always been presented as a serious and smart scholar.

Abduction – considered as the process of hypothesis generation and evaluation – plays an important role in the evaluation of arguments (whether a real shift has occurred or not): insofar as an alleged *ignoratio elenchi* is evaluated and then accepted by the audience – in a dialectical and rhetorical context – it becomes a good argument.

In the three cases of *ignoratio elenchi* illustrated above, the abductive skills involved concern the ability of turning information about people, and the social characters they represent, into relevant knowledge supporting one view rather than another. In the *argumentum ad hominem* this ability is related to the formulation of those abductive inferences, which successfully employ and evaluate the information discrediting your opponent. In the *argumentum ad verecundiam* the abductive process involved is connected to the selection of experts and authorities who may be recognized as such by a certain audience. Finally, the case of the *argumentum ad populum* involves the selection of the majority to be heeded, taking its composition into account. The possibility of gaining the favor of an audience also depends on contextual variables such as value orientations or beliefs, previous knowledge on the matter, and so on.

As argued by Dessalles [2000], logic itself could have evolved as a powerful tool able to give us reliable indicators for assessing individual linguistic competence and arguments. This contention needs some refinement. If fallacy evaluation results from an abductive process performed by the speaker as well as the listener, then we may argue that a genuine competence may have some relation with the abductive skills used by the speaker. In this sense, the success of an individual is closely related to his/her ability to display sophisticated abductive skills (or those judged as such).[10]

1.3 Gossip, Reasoning, and Knowledge

1.3.1 Ignorance and Knowledge

As a matter of fact, human beings are limited in their cognitive abilities to solve problems and make decisions. To shed light on this point, consider the so-called *ignorance problem* introduced by Gabbay and Woods [2005]. The ignorance problem can be seen as a general frame for investigating knowledge and human cognition, since most of the activities humans are involved in can be precisely described in these terms.

---

[10] Indeed, the ability to display such sophisticated abductive skills has only ever been available to a minority of human beings during a small recent fraction of the time span of human existence. On the relationship between logic and the pre-literal mind, see Harris [2009].

Formally speaking, an agent A has an ignorance problem in presence of a cognitive target T, which can be attained by a piece of knowledge K lacking at time t. Commonly, any ignorance problem is thought to have two solutions: 1) the case in which the agent solves it acquiring K, and 2) the case in which she cannot solve it, because she cannot acquire K. Gabbay and Woods [2005] pointed out that these two options do not exhaust the solution space available. Simply, this does not represent what a human agent does, and how she behaves. For they introduced a third solution according to which the attainment of T can only happen by means of conjecturing H, which, *if* true, would hit the target T.

As Gabbay and Woods brilliantly argued, in this case the attainment of T can only be *presumptive*, and the ignorance problem is merely *transformed* into something manageable, but is not *solved*. In fact, H is only a surrogate which may attain a lesser target. It is worth noting that the act of conjecturing H would not attain T, however H becomes a valuable candidate for producing new knowledge. This clearly captures one of the most fundamental features of *abduction*, which is its *ampliative* character [Magnani, 1992a; Magnani, 2001]. The ampliative character of abduction is somehow *Socratic*, as it is *ignorance-preserving*, as noted by Gabbay and Woods. Abduction is ignorance-preserving, since it would not allow us to acquire K, but only H. And, as long as we lack K, then we are forced to preserve our ignorance with respect of it.

In addition to that argued by Gabbay and Woods, we would add a fourth option to the problem, which is transformative, but it is still quite different from the third option. The main limit of Gabbay and Woods' account is that it does not address any concerns about the social (or extra-individual) dimension of human cognition. As a matter of fact, we do not live alone, and most of the time decisions are made in concert with other people. This is obvious, for instance, in the case of children whose parents provide them with most of the resources needed to fulfill their basic needs. Other people's help is still vital later in life; the degree to which people lean on each other is astonishing, as it is nearly impossible to directly experience any piece of knowledge one would need to make it.

In the following, I will describe the fourth option as the one in which we basically resort to the resources we can exhibit as part of a social group. These resources are available regardless of competence or domain-specific knowledge; they are simply resulting from the fact that human cognition would be biased towards social problem-solving[11] [Humphrey, 1976; Byrne and Whiten, 1997]. The fallacies I described in the previous section are three different examples of the same case showing how problems can be solved leaning on the social dimension of human reasoning. In the remaining part of this chapter, I will show how gossiping may turn out to be a resource for problem-solving. More specifically, I will claim the fallacies I listed above seem to have a *taste for gossip*.

---

[11] This contention will be detailed in section 5.1.

1.3.2   Reasoning through Gossiping

Gossip is commonly recognized as something rooted in the desire to harm or damage others. For instance, a dictionary definition of gossip is *idle talk* potentially threatening for one's reputation. But, is that always true?

In the last decade or so, the study of gossip has been focusing on two main lines of research. The first considers gossip as a way of exchanging information typical of human societies [Yerkovich, 1977; Ben-Séev, 1994; Baumeister *et al.*, 2004]; the second deals with its evolutionary dimension, especially in the case of the social evolution of language [Dunbar, 1996; Dunbar, 2004] and of coalitions and group management, in the framework of the multilevel selection [Wilson *et al.*, 2000]. According to this latter perspective, gossip may be defined as that kind of talk that aims at reporting stories – confirmed or not – about other absent people.

As just mentioned, gossip is a way of speaking, through which people manage and exchange information about others and their social setting [Yerkovich, 1977; Ben-Séev, 1994]. Think of how much time we devote to recounting or listening to the misadventures of others just to avoid making the same mistakes they made. For instance, gossiping about a friend of ours and his girlfriend or wife may provide us with some useful ideas to solve our own problems. As a matter of fact we are used to gossiping about everybody from friends to colleagues and acquaintances. People are interested in listening to and talking about what happens to other people.

According to Baumeister *et al.* [2004], gossiping turns out to be useful for cultural learning. Being part of a society or a group of persons (at school or work, for example) requires the individual to follow a set of rules and behavioral templates that have an external and rigid structure. Gossiping becomes a basis for learning conduct and moral rules that are embedded in the concrete stories and narratives people gossip about [Sabini and Silver, 1982]. In this sense, gossip serves the purpose of transmitting rules, norms and guidelines, and it contributes to social control [Wilson *et al.*, 2000]. It can also have a positive effect on ourselves, since gossiping increases self-esteem and diminishes anxiety. From a cognitive perspective, gossip constitutes an "extension of observational learning" and "a common stock of knowledge" which people can rely on [Baumeister *et al.*, 2004, p. 120].

Gossip has been usefully investigated by Dunbar [1996; 2004] from an evolutionary perspective. Dunbar argued that gossip resulted as an effective *bounding mechanism*, adaptively developed to tame social complexity, as our ancestors began living in larger groups.[12] Of course, living in larger groups had a dramatic effect on our ancestors' survival chances, because it permitted them to render hunting much more effective: hunting in large numbers improved self-defense, decreased risk from predators and increased the capacities for exploiting food provided in large quantities, for example. If this constituted a successful strategy to face various dangers, it did come at the cost of increasing cognitive demands. Living in large group

---

[12] Interestingly, Hill and Dunbar [2003] estimated that in contemporary Western society one's social network size averaged 124 individuals with a maximum of 153 individuals. They calculated human social network size on the exchange of Christmas cards.

imposed severe demands on our ancestors' cognitive system. For instance, it required the ability to interpret visual signs, to distinguish between friends and enemies, to recall faces and to make a coalition for promoting social exchange [Mithen, 1996].

The development of gossip as a bounding mechanism is strictly related to the origin of language. According to Dunbar [1996], language emerged as a more effective means of communication in modern human communities, and it was mainly devoted to transferring information concerning group organization and management, since it enabled humans to communicate about third-party relationships. Early forms of language prevailed over preexisting means of communication like *vocal grooming*, as hypothesized by Dunbar, because of their efficiency in transmitting knowledge and conveying information via a structured stream of sound [Pinker, 2003]. For instance, representations conveyed through words allow humans to have a valid substitute for *first-hand* experience, which can be easily passed on through narratives and stories as the case of gossip demonstrates.[13] Conversely, vocal grooming provided a very poor connection and limited information storage capacity with respect to the range of activities our ancestors faced daily.

Beyond its speculative character, Dunbar's hypothesis points to the fascinating conclusion that language could be originally shaped for managing larger group communication. In doing so, language made social exchange an essential cognitive asset for modern humans: it facilitated some fundamental activities for managing complex societies such as the negotiated division of labour and the collaboration of large non-kin groups. Accordingly, the evolutionary success of language might be primarily connected to its *gossip-enabling* features.[14]

The connection between gossip and language is made even more explicit by Wilson *et al.* [Wilson *et al.*, 2002]. They argued that gossip is a specific use of language resulting from an activity of social control. Although gossip can be used for *self-serving* purposes, they maintained it is primarily related to *group-serving* behaviors. That is, gossip is a major means of controlling self-serving behaviors and policing them. The three fallacies can be easily considered in the light of group-serving behaviors as well. For instance, discrediting your opponent – the *argumentum ad hominem* – can be considered as a means of controlling ideas and behaviors, which do not fit within the group. This can be deemed as a kind of immunization – functioning at group-level – from heterodoxy and alien ideas, as gossip in the form

---

[13] Dautenhan illustrated a speculative hypothesis called "the narrative intelligence hypothesis", in which she argued the connection between the peculiar narrative form of language (and its products, i.e., stories and narratives) and the development of social skills in taming social complexity [Dautenhan, 2001]

[14] This last consideration is contended by Dessalles [2000]. What he argues is slightly different: according to him, language did indeed spread out because of its efficiency in dealing with social complexity, but he goes further maintaining that the primary function of language is related to saying something relevant in order to access *social status*, which, in theory, is correlated with better chances of reproduction. Therefore, striving to acquire a better social status, the need to exchange social information does not explain relevance, but relevance explains gossip [Dessalles, 2000, p. 71].

of denigrating or damaging one's reputation is promptly triggered. In this case, an *ad hominem* serves to detect and eliminate a possible threat for the group orthodoxy that is still operating at group-level. An *ad hominem* has a rhetorical role as well, insofar as it is aimed at convincing an auditorium to dismiss your opponent's view. In this case, the target is to form new alliances based on one's reputation, rather than other criteria, for instance, coherence, truth, etc. The same can be argued about the *argumentum ad verecundiam*, which is specular with respect to the *argumentum ad hominem*. The appeal to an authority can have two major functions. First of all, it is group-serving in as far as it is based on the appeal to an authority that is supposed and acknowledged to be such within that group or community. Needless to say, an appeal to the Pope during a discussion at the *Annual Conference of the Italian Catholic Federation* would more easily resolve potential conflict than an appeal to the British Prime Minister. The members of the *Italian Catholic Federation* clearly recognize the Pope as an authority and this contributes to social cohesion. Secondly, the appeal to an authority can also have an important role in building new alliances resulting from irreducible conflicts among members who belong to the same group. In this case, an *ad verecundiam* has more of a rhetorical role than a dialectical one, because it is aimed at reconfiguring both the membership and partnership of two emerging and opposing groups.

The fallacies I surveyed above – *ad hominem*, *ad populum*, and *ad verecundiam* – have important connections with gossip. As already argued, gossip is *fallacy-enabling*, because it provides the arguer with additional resources in the form of narratives about other people, for instance, about who they are, what they do and did, etc. In turn, these resources may be deployed in the three arguments I am dealing with. In *ad hominem* we employ information which discredits the person (or the group of people) holding a different view from ours. This activity is closer to the common meaning of gossip, that is, a talk harming one's reputation.

A given source can be rejected by making use of malicious gossip that poisons it; in section 1.2.2 I mentioned the case of Professor Singer, about whom his opponents gossiped claiming that over the past twenty years since he had retired from the University of Virginia, he had no articles published in peer-reviewed journals, and so he should no longer be considered an authority on the matter under discussion. In the *ad verecundiam* we employ information which make us prone to accept a conclusion just because we trust the person who holds it. Personal achievements, experiences, moral qualities – in the form of stories ("he never lied [. . .] even when he was pressed to do so by the circumstances [. . .]") – may be easily transmitted, and so they furnish a basis for acknowledging and accepting a given person as an authority in a given field. In the last case of *the argumentum ad populum*, we accept a conclusion, because the majority of people do.

The dimension provided by gossip can be labeled as eco-cognitive, because it delivers information and resources, which are embedded in the social channels and, more generally, in the cognitive niche one lives in. Gossip provides the suitable eco-cognitive and eco-logical endowments to turn social information (resulting from gossip) into potentially relevant knowledge for other purposes (resolving conflicts, building new alliances, etc.). In this sense, it can be considered as a way of reasoning that I may label *gossiping*, since it makes use of social information about other people, usually obtained and transferred through gossip:[15] for example, the role a person plays in society, his achievements as well as his failures, etc. That is, the *premissory* starting points of one's argument are obtained through gossiping (I will come back to this issue in Chapter 2).

More generally, I maintain that gossiping fallacies represent *scant-resource adjustment strategies* [Gigerenzer and Selten, 2001] based on the creation of other people as social characters, used in order to obtain more information, lessen cognitive overload and solve disagreements. That is, people make sound arguments based on their knowledge about other people. Roughly speaking, we may say that it is not a matter of *what* we know, but *who* we know. Making use of the resources embedded in social channels can be viewed as a *heuristic*. A heuristic is defined as a device that can solve a class of problems in situations with limited knowledge and time [Simon, 1977]. From a cognitive perspective, a heuristic is therefore a kind of facilitator, which helps humans to manage complex tasks transforming them into simpler ones. It can eventually contribute to creating new valuable solutions in the presence of little or poor information.[16]

1.4　Concluding Remarks

In this chapter I have tried to show how some reasoning, though fallacious, can appear to be attractive and useful for beings-like-us. Taking advantage of the *agent-based* framework on logic developed by Gabbay and Woods, I argued that fallacious arguments should not be thought to be entirely bad, because they can be symptomatic of the way human cognition evolves and functions.

Assuming this derivative conception of fallacy, I have surveyed a number of fallacies, namely, the *ad hominem, ad verecundiam,* and *ad populum,* which – I posited – are symptomatic of a general feature human cognition appears to exhibit: the tendency of making use of knowledge about the others as a fundamental cognitive asset. These fallacies – I argued – share with gossip the same cognitive strategy: the exploitation of the social dimension as a source for transmitting and manipulating information. Rather then being idle talk, gossip constitutes a way of transmitting information and managing coalitions and groups, which had a pivotal role in the evolution of language and cognition. Seemingly, the "ad" arguments I illustrated can be considered as *scant-resource adjustment strategies,* which make use of information embedded in social exchange.

---

[15] Those arguments relying on "social information" are also labeled as "ethotic arguments" [Walton, 1999a]. Walton [1999a] defined "ethotic arguments" as those arguments in which "the character (ethos) of the arguer is involved". In the following, we will develop this idea by making an explicit connection with the gossip related dimension of ethotic arguments.

[16] We will come back to heuristics in Chapter 2 when dealing with bounded rationality and biased rationality.

The task set for the next Chapter will be to give a broad cognitive meaning to the various scant-resource adjustment strategies humans employ. In particular, I will explore the following question: how could we explain the simple fact that beings-like-us are fallacious and, at the same time, that they exhibit the very sophisticated reasoning responsible for the amazing ecological success of our species? Do various scant-resource adjustment strategies and the so-called "bounded rationality" account for both human failure and success?

Chapter 2
Bounded Rationality as Biased Rationality: Virtues, Vices, and Assumptions

Introduction

This chapter aims at broadening some of the issues introduced in the last part of Chapter 1. In that chapter I maintained that fallacies are bad for logic, but could be good for surviving. A fallacy can be considered as a kind of *default reasoning*, which has an adaptive value under some circumstances. As default reasoning, fallacy allows beings-like-us to solve problems or make decisions in the absence of knowledge with respect to a certain domain. Following the crowd or mom's *sayso* are amazing competence-surrogates. Fallacies are part of a kind of rationality that in the following I will call *biased rationality*.

In the first two sections of this chapter, I will illustrate the *Bounded Rationality Theory* along with its assumptions and limitations. The Bounded Rationality Theory – now becoming mainstream – is of such importance, because it furnishes the theoretical background for my proposal. As Herbert H. Simon argued, "rationality is bounded when it falls short of omniscience. And failures of omniscience are largely failures of knowing all the alternatives [...]" [Simon, 1979, p. 502]. This is precisely in line with the agent-based approach I illustrated in section 1.1. The Bounded Rationality Theory warns us to adopt a cognitive agenda, which does not seriously take into account what a person *can* actually do, and not what she *should* do.

In section 2.3, I will present biased rationality as a particular interpretation of bounded rationality. More precisely, I will illustrate the idea of *Homo Heuristicus* introduced by Gigerenzer and colleagues. In that section, I will be dealing with some issues related to heuristics and biases with relation to fallacious reasoning. More precisely, I will argue that the rationale of biased rationality consists in turning ignorance into a cognitive virtue.

In the two next sections, I will go back to the question of fallacious reasoning presented in Chapter 1. Taking advantage of the distinction between *competence-independent information* and *competence-dependent information*, I will maintain that the adaptive role of biased rationality is conditional, as they lack what I call

E. Bardone: Seeking Chances, COSMOS 13, pp. 21–46, 2011.
springerlink.com © Springer-Verlag Berlin Heidelberg 2011

symptomaticity. That is, the adaptive value of fallacies are limited to those situations in which we do not have information, whereas *being* in such a situation is not adaptive at all. In fact, As humans learn and accumulate knowledge, what will call fallacy turn out to be poorly adaptive, because they block any possible improvement due to learning.

Finally, I will broaden the problem of fallacies in the last section with respect to the process of *cognitive niche impoverishment*. This is meant to stress the fact that the resources delivered via the cognitive niche are not to be taken for granted. In doing so, I will introduce the notion of *cognitive ochlocracy* so as to give a tentative description of the phenomenon of cognitive niche impoverishment.

2.1   Laying Down the Main Assumptions of the Bounded Rationality Model

The concept of bounded rationality was first introduced by Herbert Simon in one of his first and most well-known works [Simon, 1947]. Since 1947 Simon himself returned to the point many times [Simon, 1955; Simon, 1978; Simon, 1979; Simon, 1983] and other scholars also used bounded rationality [Cyert and March, 1963; Nelson and G.Winter, 1972].

Most theories refer to BR as a powerful analytical basis, without questioning or modifying it directly [Foss, 2003; Conlisk, 1996] (for an exception to this general trend, see Selten [Selten, 1998]). For this and other reasons, that will be addressed below, I prefer to refer directly to Simon's approach [Lipman, 1995, p. 43].

As for Simon, "rationality is concerned with the selection of preferred behavior alternatives in terms of some system of values whereby the consequences of behavior can be evaluated" [Simon, 1947, p. 84]. That is to say that rationality is about: (1) the selection of alternatives through a (2) system of values that allows individuals to (3) behave in some way that can be submitted to (4) evaluation in its actual and potential consequences. Hence, rationality is concerned first with problem-solving and decision-making activities, and then with the evaluation of results.

"Alternative selection", has to do with alternative searching. To express it more precisely, nothing can be selected if we do not look for alternatives first. Thus, the process of seeking alternatives is fundamental in decision-making. It is worth noting that alternatives are to be made, i.e. they are produced by the individual engaged in the decision-making process. Since alternatives are not exogenously given, I focus my attention on two distinct possibilities. First, if the individual accesses to all possible alternatives, i.e. she/he is capable of creating a map of actual and potential effects of her/his behavior, we say she/he is fully-rational. Second, if the individual cannot produce the overall range of alternatives, i.e. she/he has limited computational capabilities and/or doesn't have full access to environmental data and variables, we say he/she is capable of bounded rationality only. The former is the economic man of the neoclassical approach describing "how people *ought* to behave, not how they do behave" [Simon, 1959, p. 254]. The latter is the real decision-maker, essentially limited in his/her computational capabilities by internal and external limits (see also [Todd and Gigerenzer, 2003]).

In other terms, to use Simon's own definition, "rationality is bounded when it falls short of omniscience". And the failures of omniscience are largely failures of knowing all the alternatives, uncertainty about relevant exogenous events, and inability to calculate consequences" [Simon, 1979, p. 502]. Broadly speaking, limits on rationality derive from natural constraints of human perception and from the fact that we are incapable of computing the overall range of possible events. The first set of constraints relate to the environment, while the second refers directly to human rational capabilities [Simon, 1955, p. 101].

Focusing on social organizations gave Simon a more detailed approach on the rationality issue. He tried to show that decision-making needs to relate with advancements in psychology and social psychology studies [Simon, 1947; Simon, 1955]. The hypothesis is that human rationality cannot map all environmental variables, create all the deriving alternatives, scan them, and then select the optimal option. In other words, if we are not able to maximize, we can only make an approximation to the optimal option. Therefore, individuals with bounded rationality reach satisficing results, i.e. they can only approximate the optimal result that is typical of the neoclassical equilibrium theories. Here the role of search mechanisms is crucial. Individual decision-making is based on seeking alternatives, and selecting them on the basis of a definite set of values. The metaphor Simon uses in order to explain this kind of searching is the decision tree [Simon, 1947; Newell and Simon, 1972], where each alternative is put a payoff. It emerges that the idea of rationality is completely related to computational capabilities [March, 1978, p. 590] rather than to the broader cognitive system. Behavior can always be defined through algorithms, even if in an imperfect way, and the bounded rational woman/man compute (the acts of searching-evaluating-selecting) which alternative could be more appropriate.[1]

The result leads then only to the *satisficing* and not to the *optimal* option. Following this approach, Simon rejects the principle of the one-best-way, introducing the concept of second best. This implies that solutions to problems or selection of alternatives can be only sub-optimal, in the real world. Moreover, we may obtain, and we normally do, more than one sub-optimal alternatives (or solutions) to a given situation (or problem). Thus, the bounded rationality model tries to take into account variety and complications in decision-making processes.

In discussing decision making processes, Simon placed great emphasis on the distinction between *substantive* and *procedural* rationality. He described that difference stating that:

---

[1] This is the case of the so called "maximizing under constraints" approach to bounded rationality [Stigler, 1961]. As Todd and Gigerenzer wrote, "[i]ntroducing real constraints does makes this approach more realistic, but maintaining the ideal of optimization, that is, calculating an optimal stopping point, does not. What is lost is psychological plausibility, because such an ideal of optimization invokes new kinds of omniscience, being able to foresee what additional information further search would bring, what it would cost, and what opportunities one would forgo during that search" [Todd and Gigerenzer, 2003, pp. 45–46].

> [...] we must give an account not only of *substantive* rationality, the extent to which appropriate courses of action are chosen, but also *procedural* rationality, the effectiveness, in light of human cognitive powers and limitations, of the procedures used to choose actions [Simon, 1978, p. 9].

According to substantive rationality, the rational character of decision-making is concerned with the result one could get following the "appropriate" actions. Whereas procedural rationality points out the procedure and the process by which people make decisions. According to Simon, bounded rationality belongs to the latter category, because it does not look only at the result one could get, but also at the way people make decisions [Simon, 1978].

In contrast, the traditional model of rationality (i.e. the neoclassical) mixes the two aspects. The model is based on variable maximization (procedure), where we obtain the only possible appropriate behavior as a result (*substantive*). Bounded rationality theory just points out that the result we can get is only a satisficing one, i.e. we get approximating solutions [Simon, 1955]. The described process is merely a computational one. When we get *approximating* results, we are supposed to make computations on external and internal variables, limited by our perceptive system and bounded rationality. If so, rationality remains a maximizing procedure, namely, the *brute force* or computational procedure that humans can only partly employ.

The result is a serious challenge to the traditional neoclassical model (or the SEU, cf. Neumann and Morgenstern [1944]) that remains, maybe, consistent in normative-prescriptive terms but completely fails in its descriptive-behavioral attempts (for a clear distinction between the two aspects, see March [1978], Frank [1988], and Etzioni [1988]). The BR model has been, and still is, a powerful concept directed to opening the "brain as a black-box" hypotheses of the traditional economic theories. Simon opens human economic reasoning to other disciplinary domains, such as psychology, social psychology, computer science, cognitive science, politics, and so on. In other terms, it was an outstanding first step towards the search for a more realistic way to define human behavior.

2.2 Getting in the Dirty: Major Constraints and Problems

My main take that I am going to elaborate in this section is that the main difficulty that bounded rationality encounters is that of explaining the overall range of successful performances [Hanoch, 2002]. In this case, the argument of bounded rationality is leaking [Beach, 1997; Beach, 1998]. While it is empirically grounded that individuals display severe computational limits, they actually carry out very complex tasks that do not simply approximate the best solution. This can be easily demonstrated using two different claims.

On the one hand, following Simon's approach, the optimum result can never be reached [Simon, 1979]. I suggest, additionally, that the optimal result cannot even be envisaged. This is a relativity-based position, and I can argue that if we do not know what the optimum is, why do we approximate? More precisely, to what do we approximate? Nonetheless, actual behavior shows successful results as well. This supports the hypothesis that successful individual behavior can be evaluated through many procedural methodologies, through which the computational remains the one related to the non-reachable and non-thinkable optimum [Mitchell and Beach, 1990].

But what about when people are successful? The history of human discoveries displays plenty of these amazing and highly successful results. How can the bounded rationality theory explain that?

I start by differentiating between procedures and results as a way in which to analyze successful outcomes and then I turn to the notion of heuristics as a possible explanation of BR uses. The case of emotion helps in defining the weaknesses of this crucial point.

2.2.1 Procedures and Results

In order to illustrate my point, let me introduce an important distinction between procedures and results. This is a common distinction in managerial science, where efficiency is the measure of the way in which results are organized, that is the procedures (or processes), while effectiveness is the measure of results on the basis of the original goal. The first refers to means, the second to goal evaluation [Simon, 1947].

In the case of the neoclassical approach to rationality, we have "optimizing"[2] procedures and "optimum" results (see Table below). Let me use these terms without directly referring to the neoclassical tradition of thought. In the first instance, optimum can be considered as an *end-state* notion [Hempel, 1966]. That is, we do not care about how to get a certain result: we just look at the outcome of a decision whether it is optimum or not (efficacy). According to that, optimum results can be regarded as the best results possible (i.e. always successful), in given conditions, i.e. *ceteris paribus*.

Secondly, optimum can also be considered as an outcome that is strictly defined by a given strategy. In this case, the term "optimum" refers to a procedure that is the optimizing strategy or "brute force" strategy (efficiency). Therefore, we may have an *end-state* optimum or optimum result (that I may call the best or always successful result) that may be independent to the optimizing strategy. To sum up: optimizing procedure is not equivalent to the optimum result or, likewise it can be said that procedural effectiveness may not coincide with the best result possible. In other terms, if we focus on results and on procedures independently, we might obtain different outcomes [Mintzberg, 1989].

The case might be that of, for example, the senior manager of a medium-sized enterprise "sensing" a great opportunity in terms of increasing the corporate market

---

[2] I employ the terms "maximizing" and "optimizing" in their slightly different meanings. In fact, maximizing is the process of reaching the maximal result, on a given set of variables; in this sense, it maintains a mathematic flavor. Optimizing is the best result, i.e. maximization of the entire spectrum of variables. While Simon's approach might be referred to the first (or to the "minimax" rule, see Newell and Simon [1972]) neoclassical Authors normally refer to the second.

share and revenues, and then taking a high risk at their own responsibility without following any orthodox decision-making process (i.e. search-evaluation-selection of alternatives). The problem concerned is that of reaching an effective result, while the problem with rational choice is that of efficiency. The two results – the first from "sixth-sense" or intuition, and the second from efficiency – can differ significantly. Now, suppose the efficient decision making process is "perfect," i.e. without transaction costs, the player fully-rational, with perfect knowledge assumption, and so on. How can you take into account variables such as creativity, intuition, personal attitudes, leadership, etc. within the decision making process? The neoclassical and Simon's models do not evaluate these aspects. In any case, decisions have to deal with multiple tradeoffs and externalities. In actual behavior – i.e. with transaction costs, unstable environmental conditions, and so on – these elements do count, and non-computable decisions are often successful. We can provide many cases on this point. However, let me just mention another one, related to financial traders and their use of so-called "technical financial analysis." This analysis is based on lines traced on graphs which describe the variations of stock prices in a given period, in order to forecast the trend and decide when to buy or sell. The tool has never been validated by scientific studies; arguments for its success lay on personal attitudes, intuition, creativity, knowledge of that market, etc. That is to say on the trader's non-computational capabilities.

	End-state	Procedural
Optimum	Best result	brute force strategy

Simon defines his satisficing result through procedures, and essentially computational ones; so that his model is affected by the same shotcomings as the neoclassical ones, from this point of view. In other words, my main claim is that Simon's theory simply fails in assuming that highly successful results, or optimum ones, can *only* be obtained by a brute force strategy, as if it is the only rational strategy. That is, Simon seems to deny the possibility of getting an optimum result without employing an optimizing procedure. Bounded rationality theory just states that the ensuing result from this strategy can only be approximating, because of human limits.

2.2.2 Explaining Successful Outcomes

Generally speaking, I claim bounded rationality theory is fairly grounded when it deals with explaining human failures; but it fails in coping with other situations that do not necessarily involve unsuccessful outcomes or biases. Hence, the question that still challenges the theory of bounded rationality is: If we could get a successful (or workable, in the sense of Beach [1998]) result using a different strategy, would the brute force rationality be exploited? Would we still be bounded in that way? How do bounds really work?

In sum, bounded rationality is based on the following assumptions:

1. maximizing strategy is the only way to get successful results;
2. the notion of bounded rationality refers to the extent to which we can employ the *brute force* strategy;
3. the notion of satisficing concerns the results that the limited humans could get by employing the brute force strategy;
4. humans can only partly employ this strategy since their limits and the complexity of the environment.

I accept (3) and (4), but I reject (1) and (2). This is due to the fact that these two assumptions cannot allow us to explain and account for all those situations in which humans successfully carry out complex tasks.

Beach [1998] presented a theory of decision making that is intended to overcome these difficulties; it is the so-called image theory. As they explain in summary, the theory is based on "three different schematic knowledge structures" (p.12) that decision makers use to "organize their thinking about decisions" (p.12). These structures are called images. The value image is related to general principles on which behavior (both individual and organizational) is based, i.e. to one's personal moral beliefs. The "trajectory image" is related to the goals (p.12) one tries to achieve; it underlines that everyone, in achieving determined objectives (real or abstract ones), generates a personal view (vision) about the future possible outcomes and everyone has hopes and fears about goal achievement. Here, emotions seem to be called into action, but nothing is explicitly mentioned in the text. This trajectory image seems to be very dynamic, in the sense that it constantly modifies, depending on the type of goal set, and on the means one has to achieve it. Depending on the type of goal and of the decision (procedural or one-shot) the image can dramatically change. Last, the strategic image concerns the plans set to get the result; concrete actions are called tactics, while the more abstract anticipations on future events are called forecasting (pp. 12-13). This image needs the latter to come into existence. This is the "hard core" of the theory, in the sense that it can be re-conducted to previous decision-making theories [Simon, 1955]. Strategy can change with the addition of information, that is to say with modifications in both environmental and internal (cognitive) variables.

The three images provide a dynamic interaction between means-end relations and human perception. The psychological background of the theory and its attempts to overcome the procedure-result divergences come out very clearly [Beach, 1997].

Despite the powerful set of analytical and empirical work around image theory, the arguments for "images" are still to be found. However, I refer to this theory as one of the best attempts to move forward from the first approaches to decision making, and my work intends to be a contribution that enriches the image theory perspective. As I will explain below, my objective is to analyze the inner core of human cognitive capabilities in order to provide theories of decision making with a powerful set of concepts that are able to frame human behavior. To this extent, my theory of decision making integrates (if not comprehends) image theory.

2.2.3   The Notion of Heuristics

In order to overcome all the difficulties related to explaining successful outcomes, Simon introduces what he called "approximating mechanisms" [Simon, 1955]. These mechanisms are the heuristics, or rules-of-thumb, and allow us to have a general picture of the problem one is facing. That is, they reduce the cognitive and computational demand to solve a problem or make a decision [Simon, 1955; Hanoch, 2002]. I claim that this is somehow an *ad hoc* explanation, because it cannot be fully integrated in the general model of bounded rationality. Some questions immediately come up: Is heuristics part of the rational process? Or is it just a trick to solve complex problems?

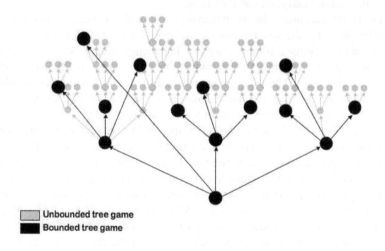

◻ Unbounded tree game
◼ Bounded tree game

My argument arises from the above mentioned distinction between procedures and results. One of the assumptions that bounded rationality theory makes is that computation (that is formalized through optimizing procedures) is the only way to get successful but satisficing results. If that is true, heuristics can only get satisficing results, because it approximately *mirrors* the optimizing procedure. It follows that the notion of heuristics itself cannot explain why humans may get successful results; it can be related to a game tree, with ex post explanations only. To make this point clearer, let me consider the Figure 1. Here, from the initial point, the decision maker can immediately reach a high alternative at the fifth level, or she/he can make the first choice and then skip to the fourth. This is, in extreme synthesis, a scheme of what we can refer to as heuristic in decision making.

It seems that the concept of heuristics refers only to a simplification of the task one faces. That is, my contention is that heuristics does not change the way we cope with a problem: it just helps to have a general picture of it. In this sense, it is something like, for example, a poorly detailed city map. One can see the wide main streets and the railway station, for instance, but not the post office, the information center, or even shops, hotels, secondary streets, and so on. Indeed, even a poorly detailed map can be useful in many cases, however, the point is that heuristics does not provide an alternative behavioral model of decision-making, because it essentially remains based on the tree game, that oversimplifies reality [Newell and Simon, 1972] were aware of these limits). In this sense, the concept of heuristics is completely coherent to an under-constrain – or bounded – procedure of optimization.

However it is, the notion of heuristics that seems to introduce a radical and different perspective about how we make decisions and solve problems. As shown above, heuristics as an *ad hoc* argument is lacking, but it can be very useful in order to devise an alternative model: it can be considered as an anomaly that we must take into account [Kahneman *et al.*, 1990]. The main reason to attribute such a role to it is that heuristics can be viewed as a *facilitator*. That is, it helps humans to manage complex tasks, and even create new valuable solutions. I will come back to this issue in section 2.3.1.

2.2.4 Emotions

In the last part of his life, Simon placed increasing emphasis on the role played by various mechanisms, such as emotions, that help bounded rationality [Simon, 1983]. Humans can be partly rational because of their limits and the complexity of the environment and, hence, they try to devise alternative paths to overcome all these constraints. This, I suggest, is connected to the idea of heuristics. As Simon put it: "emotion has particular importance because of its functions of selecting particular things in our environments as the focus of our attention" [Simon, 1983, p. 29]. Although Simon acknowledges the role of emotions in setting agenda for problem solving, he has never tried to integrate them, or other various external mechanisms, into the bounded rationality model [Hanoch, 2002].

Several authors have recently opened up new and interesting perspectives on the cognitive role played by emotions. Favored theories mainly fall into two general categories [Thagard, 2005]. The first category considers emotions as judgments about a person's general state [Oatley, 1992; Nussbaum, 2001]. Accordingly, an emotion, fear for instance, can be viewed as a result of an inference that accounts for certain clues and triggers a certain response. In this sense, emotion is a "summary appraisal" [Thagard, 2005]. Hanoch [2002], for example, pointed out that emotions, rather than being a mere constraint to rationality, are also an aid to rationality. To be more precise, they (a) "function as an information processing mechanism with their own internal logic, working in conjunction with rational calculation, [...] (b) can function as a mechanism for establishing a hierarchy of goals by pressing us to pursue goals that have high survival value while setting aside less urgent ones" (pp. 7-8), moreover, (c) emotions also let individuals imagine what can happen.

The second category considers emotions as bodily reactions. Damasio pointed out that emotions are *collections of chemical and neural responses* [Damasio, 1999, p. 51] that use the body as their theater. According to that, emotions serve two main purposes: first, the production of a certain reaction, for instance, fear may induce humans to run away, if facing danger; second, emotions regulate internal states of the organism so that it can be ready to effect a certain reaction. Increasing blood flow and breathing rhythm are examples of this kind [Damasio, 1999, p. 54].

I suggest that these two views about the nature of emotions are not alternative, but rather complementary. As stated by the first category, I may argue that emotion is a kind of representation (or cognitive state) that can be considered a part of the cognitive process involved in decision-making. It helps humans to concentrate upon what matters overcoming our limited computational resources. However, as argued in the second category, the cognitive relevance of emotions is not the one displayed by cold reasoning. In fact, emotions also involve a bodily reaction. In this sense, the representation we have of emotions is not something triggered within our conscious mind; physiological changes occur in terms of breathing rate, blood pressure, and so on. Accordingly, I may say that the body is the theatre of emotions [Damasio, 1999, p. 51]. Hence, the question is: What kind of representation is an emotion connected to? What is its nature?

Regarding emotions as both cognitive states and bodily reactions suggests an alternative model of rationality and decision-making, since emotions call for an extension of the notion of rationality used in the BR theories. The role of emotions is a kind of *opening* argument because it is something unexplained in Simon's model but, as we now see, fundamental in human behavior. More precisely, the question was: In which way do emotions affect decisions? On this point, I argued that they do it by modifying cognitive states and, at the same time, requiring bodily reactions. I now want to take things a few steps further: How do these elements affect rationality? And, furthermore, if emotions as representations are part of the cognitive process, then how about all kinds of representations? Or, more basically, what is the nature of these representations and how can they modify our model of rational choice?

2.3   Biasing Rationality

2.3.1   Introducing the Homo Heuristicus

In Chapter 1 I dealt with the problem of fallacies. My main take was that a fallacy does not necessary lead to a bad outcome. Accordingly, it can be either a good or bad argument. This is due to the fact that an argument is fallacious with relation to a standard or a set of standards. In turn, the setting of a standard depends on the resources the agent involved has. This is basically the so-called resource-based approach to fallacy introduced and popularized among logicians and cognitive scientists by John Woods and Dov Gabbay.

In the previous sections I have tackled the main issue underlying the resource-based approach to fallacy, that is, the notion of bounded rationality. I have argued that the notion of bounded rationality puts the limits of the classic models of rationality on display. In particular, bounded rationality points to the idealistic and abstract assumptions underlying the traditional models of rationality.

However, the weakest point of bounded rationality is that it does not appropriately take into account that human beings may also be successful. In order to explore this point, I will briefly illustrate the idea of *homo heuristicus* introduced and developed during the last two decades by Gigerenzer and colleagues (cf. [Gigerenzer and Goldstein, 1996; Gigerenzer and Selten, 2001; Todd and Gigerenzer, 2003; Gigerenzer, 2000; Gigerenzer and Brighton, 2009]). The idea of *homo heuristicus* explicitly addresses the problem of how to make two apparently conflicting concepts consistent: *accuracy* as the result of a certain decision, and *effort* as the amount of resources deployed in the decision-making process.

The idea of *homo heuristicus* stems from the rejection of two main assumptions about accuracy and effort. The first is that a heuristic always involves a trade-off to be reached between accuracy and effort, as they are basically conflicting concepts. In fact, accuracy usually involves time and resources. Therefore, given the fact that humans operate in cognitive economy with limited time and resources, they have to rely on decisions that are accurate enough, meaning that they might simply have to discard those strategies which lead to more accurate outcomes, but require greater resources. Heuristics are thought to be strategies reaching an accuracy-effort trade-off (cf. section 2.2.3).

The second assumption can be called the "principle of total evidence". The principle of total evidence – introduced by Carnap [1947] and explicitly mentioned by Gigerenzer and colleagues – states that it is always better to take into account the total evidence available in order to determine whether or not a certain hypothesis or course of action is justified or rational: that is, having more information is always better than having less information. Or, To put it simply, *more is always more*, and *less is always less*.

Contrary to these two beliefs, Gigerenzer and colleagues argued, and managed to provide empirical evidence to support the idea, that heuristics are not always accuracy-effort trade-offs. On certain occasions, one can attain higher accuracy with less effort. Besides, more information may be detrimental leading not only to overload, but also to a general state of ignorance. Putting it simply, *less is more* and *more is less*.

An example illustrating this point is the so-called "recognition-heuristic". What Gigerenzer and his team found is that when facing two alternatives, the one that is recognized is usually selected [Raab and Gigerenzer, 2005]. In an interesting study, Raab and Gigerenzer asked two groups of university students respectively a German one and an American one, which city has a larger population between San Diego and San Antonio. Quite surprisingly, 100% of German students responded correctly, whereas only two thirds of American Students got the answer right. How could that be possible? We would expect American students to get it right, as San Diego and San Antonio are two American cities, and therefore they should know more about them or, at least, have more information. The explanation provided by Gigerenzer and colleagues is quite cunning. German students got it right, because they know less. More precisely, they got it right, because they only recognized one of the two cities, and thus they thought that it should have been the largest between the two. This is a fair example of recognition heuristic.

Gigerenzer and colleagues studied and tested a number of heuristics that turn out to be smart strategies for solving problems or making decisions. These heuristics compose what is called the "adaptive toolbox" [Gigerenzer, 2000; Gigerenzer and Brighton, 2009]. Basically, this is a set of *fast-and-frugal* strategies that allows us to attain high accuracy while still operating in cognitive economy. For instance, *fluency heuristic*. In the fluency heuristic, both the of the two alternatives are recognized, but the one that is recognized faster is picked. So, going back to the example of the two cities, if I recognize San Diego faster than San Antonio, then I will choose it. Another example of smart heuristic is the so-called take the best. The best way to choose among concurrent and recognized options is to search for clues and stop as soon as one finds a discriminating clue favoring one above the others.

The idea of *less is more* is less paradoxical than one might think. It means that we simply operate selections: we basically search for the information that we think is relevant. Indeed, the selections we make are to some extent arbitrary, since we know what we select, but we do not know what we leave out. Heuristics are strategies – partly dependent on the context in which one is operating – that serve to this purpose: making use of what we know. In order to detail this point let me come back to the problem of fallacies. I will introduce and illustrate an alternative framework, which does not contrast with the results popularized by the proponents of the adaptive toolbox. It will however be useful for incorporating the idea of *less is more* into a broader picture.

2.3.2 Easy to Use: The Rationale of Biased Rationality

The very idea behind the *homo heuristicus* furnishes an account about why biased minds make better inferences. In fact, heuristics are biases. Traditionally speaking, biases have been always considered as psychologically complex, leading to negative or unhappy outcomes. A bias is not necessarily an error, but it is usually considered as resulting from a poor or lower form of rationality, namely, biased rationality. They can indeed speed up a decision-making process, but, generally speaking, they are not necessarily a response to cognitive economy: they are easy to deploy them.

In his *Irrationality. The Enemy Within* [2000] Sutherland listed a number of biases that are, at the same time, errors and successful cognitive strategies, just as Gigerenzer and colleagues showed. Let me mention a few of them, namely, the *availability bias*, the *primacy bias*, and the *halo effect*. I mention them here, because they have something in common with those heuristics I have just dealt with.

The *availability bias* means seizing the first impression one comes up with about a person, an object or a situation. For instance, after witnessing a car crash, drivers inevitably tend to slow down, because the scene brings to mind the possibility of being involved in a similar accident.

The *primacy bias* consists in interpreting certain clues in light of those presented earlier. This means that we usually continue to make use of first impressions causing a sort of priming effect. For instance, a teacher may favor certain students over others simply because they obtained good marks in earlier examinations. Basically, biased teachers would tend to color the interpretation of the students' subsequent performance according to the earlier result.

The *halo effect* is a further specification of the *availability bias* and the *primacy bias*. Basically, it occurs when a person judges a situation, an object or another person relying only on one good trait. An example is a script presented in good handwriting. On average, scripts in good handwriting are rated higher than those which are badly presented. In this case, good handwriting is the good trait one uses in order formulate judgment overlooking other factors that might cause a change in judgment.

These three biases are commonly considered as errors [Sutherland, 2000]. But what kind of error? One can be biased, but at the same time achieve a good performance, just like in the cases illustrated by Gigerenzer and colleagues. For instance, a teacher may be biased towards certain students and those students might still perform very well at the same time. In this sense, it is not an error to get things right when judging on first impressions or coloring the interpretation of students' later performance based on initial results. It might however be an error because, if we got things wrong, we could be more easily blamed for our mistake and we would be told not to rush to judgment. We attribute a low cognitive status to biases because it is easy to highlight their weaker points, even though their weakest points do not necessarily lead us to be mistaken. It is this sort of *blamability* – actual or potential, it does not matter – that is the source of mistrust. Basically, we have a commitment towards negotiating "the journey from cognitively virtuous starting points to cognitively virtuous outcomes" [Woods, 2009]: that is, we start out safely and we want to arrive safely. Blamability warns us of the fact that we did not take the wrong path, but that it was a dangerous one.

More generally, I maintain that this attitude rests on the human capacity for planning ahead. Basically, when confronting a problem, people try to foresee possible objections, usually taking some precautions. Depending on their abilities and skills, people may anticipate some of the negative consequences a certain course of action might have. Planning ahead is somehow a certification that unhappy consequences may be prevented, even though we do not know precisely whether they are going to happen or not. Therefore, what we consider erroneous is the way biases (but also fallacies, as we will see) manage possible objections to a decision and/or unhappy consequences.

Let me now go back the question about which city has a larger population between San Diego and San Antonio. As already mentioned, the German group answered correctly, because they did not know San Antonio. Indeed, insofar as they were not mistaken, we could not blame them because for their ignorance. However, in the long run ignorance might be a problem.

2.3.3 Appealing to Ignorance and Its Cognitive Virtue

In order to clarify this point, I connect the recognition heuristic with an argument – traditionally deemed as fallacious – the so called *argumentum ad ignorantiam*. Let me make a very simple example. Suppose that John has to attend a meeting in the afternoon at his department, but he has not received any communication yet. Usually, department meetings are announced at least a few days before by the head of the department who sends an email to all the staff members. But this time she did not send any email to her colleagues. The meeting would usually start in less than one hour and John does not know what to do. Then, he carries out the following reasoning:

1. If there were a meeting at my department I would know it.
2. I do not know such a thing.
3. Thus there will not be a meeting today at my department.

This is considered by traditionally-minded fallacy theorists to be a fallacy. The main reason is that ignorance is never probative, meaning that it can only prove that one does not know a thing. In fact, in my example there might be a number of reasons why John did not come to know whether the meeting was going to happen or not. 1) Maybe his colleague sent him an email he never received, perhaps the head of department did not type his email address correctly, and then forgot to pay attention to the delivery failure message she should have received back. Carelessness in this case could be the reason explaining his ignorance. 2) Maybe his colleagues did not inform him about the meeting because they noticed he had not shown up during the last week, and they thought he had taken a week off to work on the last chapter of his book. Or 3) the head of the department and his other colleagues did not inform him on purpose, because they wanted to mob him.

In this case, we do not need to know whether John is eventually right or wrong but we can immediately see how it is easy – from an intellectual perspective – to raise some objections to John's argument. All the objections we were pointing to are related to the fact that John could have relied on a better argument.

John followed a pattern of reasoning that is labeled by AI theorists as *autoepistemic reasoning*. As Gabbay and Woods put it, "autoepistemic inferences are presumptive in character". Given that a candidate hypothesis is not known to be true, it is presumed to be untrue" [Gabbay and Woods, 2005]. It is also known as negation as failure [Walton, 1995] or *argumentum ad ignorantiam*. An *ad ignorantiam* consists in an explicit appeal to our ignorance. In general, I analytically describe it as follows:

1. John knows he does not know P.
2. John asks himself whether he would have known P.
3. He would have known P, if it had been true.
4. He does not know P.
5. Then, he knows P is false.

What is interesting about this formulation is that it stresses how we are able to turn our ignorance into a cognitive virtue generating premissory starting points that we previously lacked. In fact, In 1) P is what prevents John from deciding. Conversely, in 4) P now becomes a clue suggesting a possible conclusion. It is 3) that describes the move allowing to escape ignorance without overcoming it. This ignorance-escaping feature of an *argumentum ad ignorantiam* should be treated along with another one: the move described in 3) is *ignorance-escaping* insofar as it is *irrelevance-avoiding*. In 1) John lacks premissory starting points that are relevant to the matter. In fact, in my brief debunking we brought up a number of objections that explicitly called for relevant information that should be acquired beforehand.

However, although John does not overcome his ignorance by acquiring new and relevant facts, he escapes from it by avoiding any commitment to being relevant. That is, an *ad ignorantiam* does not get us out of ignorance, but it makes unapparent the distinction between relevant information and irrelevant information (I will be back to this problem in section 2.4.1 when dealing with the epistemic bubble introduced in Chapter 1).

Relevancy avoidance is successfully performed, because it permits people to make a decision no matter what they know. The recognition heuristic chosen by the German group – just like in the case of *ad ignorantiam* – makes use of the same pattern of reasoning: it takes the lack of recognition as *probative*. The implicit passage in this case is as follows:

1. If I had known San Antonio, it would have been larger than San Diego.
2. But I do not know San Antonio.
3. Therefore, it is smaller than San Diego.

One important thing should be specified. The option generated by making use of our ignorance is not like the one we would gain by flipping a coin. It is certainly less arbitrary and more sophisticated, because it is at least a *spin*, as Woods argued. In fact, it permits us to make some guesswork possible. That is, we are not wholly in the dark. In fact, as already mentioned, an *ad ignorance* permits us to unfold premissory starting points, as they at least make a certain decision decidable or affordable.

Going back to our example, it is not true that John does not know anything. For instance, he knows something about what he should know. So, he can easily withdraw the hypothesis that he has not been informed because the head of the department wrongly typed his email address, because she always sets a return receipt option for such emails. She would let John know in the case she did not receive any confirmation from him. He could easily withdraw the second objection, because he usually works from home. And, as for the third one, he could discard that as well, because he has no problems at all with his colleagues.

The conclusion we are now arriving at is that an *ad ignorantiam* – belonging to biased rationality – is a weaker cognitive strategy than the one relying on relevant information.

2.4 The Vices of Biased Rationality

In the last section I discussed the idea of bounded rationality as biased rationality. I made an explicit connection with the idea of *homo heuristicus* and his potentially fallacious dimension. In this section I will illustrate the problem of biased rationality going back to discuss fallacious reasoning with relation to the problem of relevancy. The treatment of this issue will be a crucial cornerstone in the introduction of the proposal to be developed in the following chapters.

2.4.1   Competence-Dependent Information and Competence-Independent Information

In Chapter 1 I looked in detail at number of fallacies identified as *ignoratio elenchi* (or *red herring*). By definition this class of fallacies introduces irrelevant information. The *argumentum ad ignorance* too can be classified as an *ignoratio elenchi*, as I have just discussed. In section 1.2.4 I pointed out that irrelevancy is always a relative matter, as it also depends on the communicative context a certain reasoning is involved in. In this section I will try to make a step forward re-addressing the matter within a broader framework. This would also allow us to go beyond biased rationality (cf. section 2.5) and then to illustrate the concept of biased rationality as cognitive ochlocracy in section 2.6.

My main contention is that *ignoratio elenchi* is a kind of argument based on the introduction of what I call *competence-independent information*. An *ignoratio elenchi* is selected when a person does not have those competencies allowing him/her to address the original issue for debate. This kind of strategy may tell us something insightful about the nature of human decision-making and their cognitive system.

Consider for example a problem which many people face during their holiday: when to go swimming after eating. A doctor explains that there are several variables which we should account for in order to decide when to go swimming after eating. It depends on how much we eat, what we drink, the water temperature, whether we swim hard or not. All this information is *relevant* when deciding what to do and when to do it. Why is it relevant? It is relevant, because it would *explain* whether we may get cramps or other problems related to digestion. For instance, a heavy meal eaten just before swimming would make you feel sluggish and thus *explain* cramps.

As many people do not have the competencies a doctor has (or is supposed to have), they often rely on other kinds of information. For instance, mom's suggestions or what the majority do. By definition, mom's sayso or what the majority of people do are all irrelevant information. In my example, this information is irrelevant because it would not explain whether we may get cramps or not. More precisely, it is irrelevant because it is not *symptomatic*: what other people decide does not explain why a heavy meal affects our metabolism making us feel queasy.

The introduction and adoption of irrelevant information can be motivated by various reasoning. Indeed, it may be a strategy to divert audience attention and thus challenge the original issue for debate (cf. section 1.2.4). Think for instance of how often politicians attack their opponents personally, not their ideas or the opinions they hold. However, as far as I am concerned here, I maintain that the introduction of irrelevant information is primarily a *cognitive strategy*, which responds to the necessity of *cognitive economy*. More precisely, it is a strategy that is deployed *in the absence of competence* regarding the matter in discussion. Thus, competence is connected with relevancy: being competent with regards to a certain matter is what permits employment of information which is relevant or, more precisely, symptomatic.

Focusing on competence so defined may help us solve or, at least, explore some open questions related to fallacy and biased rationality. First of all, what I argue is that the fallacious nature of biased rationality is concerned with the introduction of information that is not symptomatic to the conclusion that is drawn. Your mom's suggestions do not explain why you may have cramps or not; whereas what you have eaten does. The distinction between the two kinds of information can also be described in abductive terms: competence-dependent information are those clues or signs "from which we can infer that a given fact must have been seen" [Peirce, 1931–1958].[3] In my view, irrelevancy is therefore an *epistemological* feature of fallacious reasoning and biased rationality and, above all, it characterizes them. Since the information we provide is not symptomatic, it is always *irrelevant*. It is not the case that it is *sometimes relevant, sometimes irrelevant* depending on the context. This information is always irrelevant, because it is not symptomatic.

The question about competence and symptomaticity may also clarify the reason why a fallacy or bias is a *sometimes good, sometimes bad* strategy. This was already clearly recognized by Aristotle who extensively argued around the unapparent defectiveness of fallacious reasoning. More precisely, the particular feature of fallacy is that it appears to have a certain property, when it has not. This unapparent defectiveness is connected with the fact that the information deployed in some fallacious reasoning is not symptomatic. As already pointed out, fallacious reasoning does not explain the reason why a certain event is *such and such*, and not *as such as such*. However, even if fallacies are not symptomatic, they can lead us to solve the task we are supposed to face, as already pointed out in the last section. Irrelevancy – intended as not-symptomatic – is what we fail to be aware of, because in that case:

1. we do not have any other resources to deploy, and thus we would not make it;
2. symptomaticity is not a prerequisite for target attainment.

The first point is concerned with the relativity of error we have already dealt with in section 1.1.2. An argument turns out to be fallacious – and thus biased rationality – with relation to a target which embeds a standard it fails to meet. This is basically a conception of error, which is related to the selection of standards that overcomes the agent's capabilities and resources. Here, particular attention is given to the fact that for *beings-like-us* some choices or decisions are to be made independently of the resources or capabilities that one may or not have. This can be considered a phenomenon of *cognitive immunization*. According to Gabbay and Woods [2005], cognitive immunization is that kind of impediment from becoming aware of the phenomenological structure of the *epistemic bubble* we are in (cf. section 1.1.1). By definition being in an epistemic bubble makes phenomenologically unapparent the distinction between *knowing* that p and *believing* to know that p.

In my view, the notion of cognitive immunization may be fruitfully extended to the situation that prevents us from being (or becoming) aware that we would not have those competencies allowing us to make it. More precisely, I argue that the process of embubblement also regards the kind of information we use to solve problems so that in any epistemic bubble some information appears to be symptomatic, although it is not. In the example illustrated above, we consider trusting mom's advice or following the crowd as it could be symptomatic with digestion, but clearly it

---

[3] The notion of abduction will be more extensively illustrated in section 4.3.

is not. The cognitive immunization process permits us to sort out, and make use of information that we already have, even though it is not relevant.

I claim that the process of embubblement is, indeed, essential to the deployment of biased rationality as an adaptive tool. In fact, the importance of having such a mechanism of immunization rests on the *mandatory character* of certain decisions, which have a direct or indirect connection to our survival and reproduction. The example introduced above shows how we can solve a task without having any of those competencies required to make a sound judgment. We simply rely on forms of reasoning which provide information that are not dependent on any particular competence. In sum:

- it is some kind of *easy-to-deploy* or *default* reasoning. Since it does not require particular knowledge or competence, it is a kind of default reasoning as it is easily available to everybody;
- it allows us to *avoid irrelevancy*. No matter if it introduces irrelevant information or not, it may allow us to accomplish the task we have to face trapping us in an *epistemic bubble*. This particular feature also permits us to escape from ignorance by avoiding relevancy, as we have pointed out the previous section;
- it is *domain-independent*; since it is not based on particular body of knowledge, it can be employed with regards to a number of different issues. For instance, I can trust my mother and take her advice, whether it is about how to dress, how to eat, the university to go to, the job to choose and so on. In the same way, we can follow the reasoning of the crowd in choosing what to buy, the restaurant which to go to, etc.;
- it is *resource-saving* or a*cognitive thirst quencher*. It does not base its attractiveness or appeal on the fact that it provides additional resources to solve a problem. Quite the contrary, it bases its appeal on quenching our thirst for information and cognitive resources.

This last point is clearly captured by the principle of *knowledge debased* introduced by Woods who wrote:

> Since we survive and prosper and sometimes build great civilizations in the absence of knowledge, that is to say, under conditions of widespread ignorance, knowledge is of no essential value to these achievements. [Woods, 2009]

2.4.2 Having Poor Information and Having No Information at All

It is worth noting that limits have more to do with competence-dependent information than competence-independent information. Since the latter is not strictly symptomatic about whether a certain situation is going to happen or not, we are not expected to modify it to enhance our performance. This is the problem of bounded rationality as biased rationality.

The strategies based on competence-dependent information acquire an adaptive value, as they supply cognitive resources that are much more reliable than the ones based on competence-independent information. For instance, in the experiment about which city is larger between San Diego and San Antonio, it is most likely that an American expert in urban studies would not rely on recognition heuristics. More generally, experts tend to be *de-biased*, so to say. This is so, because knowledge – and consequently the strategies based on it – is an increasingly reliable means for solving problems the more *abundant* it becomes. This is basically derived from the fact that knowledge is *resource-consuming*, whereas fallacies and biases are *resource-saving*. Strictly speaking, they are not corrigible, because they are optimized and ready to do their job. In this sense, arguments leaning on competence-independent information are neither corrigible nor enhancing.

This last contention is connected with a point introduced in the previous section, namely that fallacies and biases are somehow easy to dismiss. For instance, when people say that someone is biased or that an argument is fallacious they do not really mean that the person is mistaken or that the line of argument is wrong or false. For example, when during a debate or discussion a person claims that the opponent is diverting audience attention by performing a personal attack, they are not claiming that what their opponent argues is right or wrong. That would be a fallacy, namely, the so-called *argument from fallacy*. Instead, they are merely pointing to a flaw in their opponent's reasoning, which might eventually lead to a bad decision or outcome, if followed.

So, being fallacious or biased renders an argument easy to dismiss. That one can be easily dismissed is not to be intended as a logical derivation. However weak or easy to dismiss an argument or bias is, it may allow a person to reach his target, as already maintained. Conversely, it describes a communicative move. For example, the simplest case of an *ad hominem* could help us make a decision which may eventually be a good decision. Sometimes a person who has, for instance, a conflict of interest may be biased in holding certain positions. Therefore, knowing that he has a conflict of interest is not irrelevant at all.

More generally, my point is that irrelevance is a communicative feature. Irrelevance simply warns us that a certain piece of information may support an *easy-to-dismiss* point. Therefore, it prompts us to change or adjust our argument in order to acquire a better chance to succeed in a given discussion.

2.5 Appealing to Knowledge: De-biasing Rationality

So far I have discussed the fact that bounded rationality as biased rationality is indeed a survival strategy, however ill-grounded it may be from a more sophisticated perspective, namely, an intellectual one. The fact that we recognize some arguments or strategies as easy-to-dismiss simply means that we could have some better arguments and strategies at our disposal do we really have some better strategies or arguments? And what happens if we continue to use easy-to-dismiss strategies, even when they are not the only solution available? The following will attempt to set the stage for a possible answer to the first question. Basically, I will argue that it is not necessary to reject the idea of bounded rationality to dismiss that of biased rationality. Whereas in section 2.6 I will provide an answer to the second question. More precisely, I will illustrate when biased rationality may lead to what I call "cognitive ochlocracy" as a detrimental form of cognitive resource management.

2.5.1   Plastic Behaviors and the Lens Model

As already pointed out, fallacies and heuristics make some decisions affordable for us, even though they do not resort to symptomatic information. This is due to the fact that biased rationality, as a set of fallacies and heuristics for decision-making, turns our ignorance into a cognitive virtue generating premissory starting points. So far so good. However, we do not overcome our ignorance, as we come up with premises that are not symptomatic or, at least, ambiguous. We overcome ignorance by establishing procedures that deliver information or resources that are somehow relevant or not biased. This can only happen by building up external structures that provide us with clues or information that are more symptomatic than the ones we previously had. So, I introduce an eco-cognitive element, which will be the cornerstone of the argument I will develop in the next chapter introducing the idea of de-biasing rationality as distributing cognition.

Human beings owe their ecological dominance to their ability to display advanced plastic behaviors. In turn, the possibility to display advanced plastic behaviors is closely related to having a second, non genetic, source of information, which upon occasion can deliver the proper resources to solve problems and help make decisions. Ultimately the ability to turn available raw materials into cognitive resources to support plastic responses is central to human success. The lens model theory introduced by Egon Brunswik [1952; 1955; 1943] sheds some light on the dark side of biased rationality.

According to Brunswik, the relationship between the organism and the environment is defined by what he called "the lens model". The lens model is based on the idea that the relationship between the organism and the environment is mediated by the use of the so-called *proximal stimuli*, from which the organism can infer the distal state of the environment, which brought it about. *Ecological validity* is the term introduced by Brunswik to refer to the situation in which a given proximal stimulus acts as a valuable indicator of a certain distal state or event; ecological validity is a normative measure about *how diagnostic* certain proximal stimuli are with respect to a given distal event [Vicente, 2003].

The main idea behind Brunswik's lens model is that it provides an alternative way to look into the questions related to domain-independent versus domain-specific approaches popularized by evolutionary psychology. His main contribution to this issue is to distinguish between the cognitive *process* of a certain activity and its *content*. He pointed out that the cognitive process of inferring a distal state of the environment from the proximal stimuli we received is *domain-independent*. Conversely, what is domain-dependent are those *indicators* or *local representatives* we make use of in order to infer distal states of the environment. For the indicator content is left unspecified.

In the light of Brunswik's lens model, adaptation (and thus the possibility of survival and reproduction) is the degree to which an organism attains a *stable relationship* with the external world [Kirlik, 2001, p. 238]. In other words, achieving a stable relationship with the external world depends on developing *prepared* associations between a proximal stimulus and the corresponding distant event. Within this framework, plasticity is defined as that ability to change or adjust a pre-wired response according to the environment so as to increase the chance of a match between proximal stimuli and the distant state in an ever-changing environment.

As far as I am concerned here, plasticity can be defined as the ability to make use of those signs or clues that are more symptomatic of a certain event or situation than others. Ultimately, plasticity deals with the development of the abductive skills, which allow us to detect clues and use them as indicators or local representatives of a distant event. In turn, these abductive skills basically rely on knowledge and competence.[4] The crucial point for exploiting cognitive plasticity is to detect – and sometimes even create – various indicators specific to certain domains and not others in order to increase our chances of making successful inferences and judgments [Hammond and Steward, 2001].

This last contention leads us to consider cognitive strategies leaning on competence-independent information as ill-grounded for a long-term strategy insofar as it employs resources that by definition are not symptomatic. In Brunswikian terms, we may argue that the ecological validity of competence-independent information is quite poor, because it uses indicators that are not specific to a particular domain. Independence from a specific domain of application turns out to be the major limitation in this case. Mom's suggestions or following what the majority think are clues that cannot however be taken as reliable indicators or proximal stimuli of specific distal events.

2.5.2 Competence-Dependent Information Is Ecologically Delivered

In the previous discussion, I argued that some problems related to reproduction and survival are mandatory. One cannot cast them off, because that would impede reproduction and/or survival. Roughly speaking, under such conditions, giving an answer — even at random – is as good as giving the right answer. I posit that the virtue of strategies based on competence-independent information can only be *conditional*. That is, the *use* of competence-independent information in the situation where we have no information at all is "good", whereas *being* in that situation is not.

If so, one might expect that evolution would have provided human beings with a mechanism to escape from such conditions of having no information at all. My contention is that such a mechanism is not provided at an individual level, but at the *eco-cognitive* one. The strategies based on competence-dependent information are adaptive as far as knowledge can persist and be accumulated and transmitted from generation to generation via the *cognitive niche*.[5] As already pointed out, when

---

[4] I will be dealing with this point more in detail in section 4.3.

[5] The notion of the cognitive niche will be illustrated in section 3.3.

knowledge is easily available, strategies based on competence may become dominant: easily available cognitive niches make abundance of knowledge possible.

As I will point out in the next chapter, individual agents spend part of their time tending to enhancement of cognitive assets if this makes the achievement of cognitive goals possible where they were previously unaffordable or unattainable [Magnani, 2007a]. My claim is that this can only happen at the eco-cognitive level. Basically, a cognitive niche provides humans with an additional source of information storage and computational abilities, which support and even boost the capacity of exhibiting an increasingly flexible, adaptive response to an ever-changing environment. These extra-genetic materials, properly exploited by ontogenetic mechanisms like learning, provide the unique framework for re-adjusting and refining our cognitive assets *also* as individual agents.

2.6   When Biased Rationality Is Cognitive Ochlocracy

Mom's suggestions or what the majority thinks do not entirely lose their appeal. In Brunswikian terms, domain-independent indicators are replaced by domain-dependent ones, but this replacement is not permanent, since the resources which are ecologically delivered via the cognitive niche do not rely on any genetic mechanism granting their persistence *inter-* or *intra*-generation. Cognitive niches may be exposed to a process of *impoverishment*, since their maintenance is not secured by some genetic mechanism; conversely, it is administered by human communities and societies, and thus wide open even to the most dramatic changes (cf. section 3.5). The modifications brought about by cognitive niche construction are always reversible and they can also disappear in a relatively short space of time.

2.6.1   The Case of the Bandwagon

In this subsection my aim is to describe how the so-called bandwagon effect may cause a cognitive niche to be impoverished. The product of an *ad populum* (cf. section 1.2.3), the bandwagon effect is that process in which a person follows what the majority of people does [Leibenstein, 1950]. Generally speaking, popularity is a kind of *competence surrogate*. It is a competence (or knowledge) surrogate, because it does not require a learning process which is usually the only means by which competence might eventually be acquired. However, as brilliantly put it by Sunstein "conformity is often a rational course of action, but when all or most of us conform, society can end up making large mistakes" [Sunstein, 2005a, p. 3]. This is my starting point for asking some questions: what happens when the probability that a person chooses what other people choose increases? What are the cognitive consequences of that situation?

As already pointed out in subsection 2.5.1, plasticity is the ability of exploiting those signs that are more symptomatic than others of a certain situation we have to cope with. Indeed, this ability is enhanced by having a second non-genetic inheritance system, by which successful solutions can be transmitted and accumulated leading to better adaptation. This allows people to have access to a great variety of information and resources resulting from the activity of previous generations. My contention is that the bandwagon effect constitutes a major factor in cognitive niche impoverishment, because it slows down, or even interrupts, the accumulation and transmission of knowledge. Investigating citation behavior in science may be a case in point when illustrating this contention.

During the last thirty years a number of contributions about why and how authors cite each others' work has appeared [Bornmann and Daniel, 2006]. Such contributions have developed an alternative approach to sociology of science [Small, 2004; Van der Veer Martens and Goodrum, 2006] taking advantage of the introduction of powerful tracking tools able to store and retrieve upon request a huge amount of data. For citation analysis has now become a widespread methodology, which is a fundamental means for extracting meaningful patterns from various citation-databases. Indeed, there are a number of issues concerning the scientific reliability of citation analysis, but as far as I am concerned here, such data and results are merely *exploratory* rather than validating.

An interesting case to mention is given in Anderson [Anderson, 2006]. Anderson investigated the influence that a book, *The Social Psychology of Organizing* by Karl Weick, has had during the last thirty years on a number of disciplines including social psychology, management, and organizational behavior. Using a methodology called *context citation analysis*, Anderson reported empirical evidence regarding the fact that authors citing *Organizing* appear to be "willing to accept concepts in *Organizing* without empirical confirmation" (p. 1687).

This appears to contradict the commonly accepted view according to which science is built on *organized skepticism* [Merton, 1996]. In my view, the example I documented is simply due to the bandwagon effect. That is, scholars often cite documents merely because they see them cited by others. As a result, most of these citations may reasonably be considered as *perfunctory* or *ceremonial* citations rather than meaningful, as they usually appear in introductory sections [Case and Higgins, 2000]. On some occasions, authors may not even read the materials they cite, but often they merely copy them from a third source (cf. with "the Matthew Effect" in section 1.2.1).

2.6.2 The Two Main Consequences of Cognitive Ochlocracy

I maintain that the bandwagon effect can be broken down into to two main sub-effects. First of all, as time passes, some concepts and theories simply go unquestioned and unchecked. Secondly, due their fame, certain concepts or ideas actually become less well known, in detail, than one might imagine, notwithstanding their popularity.

On the first point, the bandwagon effect makes people suppress any private information as they conform to what the majority thinks [Sunstein, 2005a]. A scientist, for instance, may opt to support a particular theory, although he may have good reason not to, just because his community strongly supports it. He may deliberately choose to reject as erroneous the results he got from an experiment, because they do not conform to the view held by the majority of his colleagues. Indeed, lacking confidence may promote conformity, but, as the bandwagon effect becomes more influential, the price one has to pay for dissent may be very high.

This phenomenon may be considered as a kind *reverse* epistemic bubble – so to speak. I have already treated the problem of *embubblement* in section 2.4.1. By definition in an epistemic bubble, a person believes she knows *P*, when she does not. In a reverse epistemic bubble it happens quite the contrary:

an agent *A* believes she does not know *P*, when she does know *P*.

We are still in the presence of an embubblement process, since the difference between *believing* and *knowing* is apparently suppressed. However, in this case the suppression is somehow reversed, that is, it is about something that we would know. The effect of conformity is a fair example of this kind as the bandwagon effect may cause a person not to trust the information or knowledge she has. When the bandwagon effect is particularly strong, then reverse embubblement can easily take place. Fear or low level self-esteem may indeed boost it [Klucharev *et al.*, 2009].

Adhering to a certain view for group cohesion turns out to be damaging for the group itself, since in the long run it impairs group performance. This appears to be a paradox, but it is not. As people follow the so-called "wisdom of the crowd", the bandwagon has the cognitive effect of diminishing the total level of information available to the group [Sunstein, 2005b; Sunstein, 2007]. If at an individual level conformity allows people to make a decision when lacking competence and knowledge, at a group level this could be catastrophic especially when facing change and/or difficulties.

Most of what we think or believe results from *second-hand* knowledge. As a matter of fact, we try to learn from other people due to the exceedingly high cost of individual learning. Following the crowd would appear to be pro-social behavior underlying or even promoting such an activity like learning. However, even if the bandwagon effect certainly has a pro-social component, it cannot be smuggled in as a tendency enhancing any process of social learning. It is quite the contrary, as the bandwagon effect reduces the total level of information available within a group or a community and it drastically weakens social learning, increasing the cost related to individual learning. Conformity enhances the probability that a given behavior or trait will become common in a group or population [Efferson *et al.*, 2008]. In doing so, it reduces eco-cognitive variation within a group and consequently makes learning by imitation less profitable. In fact, as reported by Castro *et al.* [2004] imitators "do poorly when they are common and individual learners are rare".

Conformity also impedes creativity, because dissent may be of great value when innovation is required to adapt to new situations for which established and popular solutions are no longer useful. As a matter of fact, popularity makes people conform to certain cultural variants rather than others. However, in the case of strong discontinuity with the past, the attitude of conforming to what other people think is strongly maladaptive insofar as it stifles dissent, which innovative and creative solutions very often come from. In fact, innovation requires tempting alternative ways of thinking, which, however, can be impaired by an overwhelming conformity in promoting or just passing on what is already known.

There is another major consequence due to the bandwagon effect, which is worth discussing. As already mentioned in analyzing citation behavior, authors often cite well-established ideas or concepts just because they are popular. They are called ceremonial citations. However, as an idea or concept becomes popular, it is more likely to go unchecked, and debate over it dies down. This may promote a kind of *obliteration-inducing* process according to which the more a concept becomes popular, the less it is really known and understood.

This contention is indeed highly hypothetical, since it is hard to assess whether or not a scholar does understand what he cited. This phenomenon may drastically vary in breadth depending on the matter at discussion, whether it is a piece of science, art, philosophy, or religion. However, it is common to find contributions in science as well as in other disciplines which explicitly address the problem of getting back to basics as if the original concepts and/or ideas were lost.

In abductive terms, this obliteration-inducing process can be defined as *premise obliterating*. Peirce brilliantly noted that, as we get familiar with certain reasoning, a habit is established so that we tend "to obliterate all recognition of the uninteresting and complex premises from which it [a conclusion] was derived". [6] I argue popularity is a major factor producing another kind of premises obliteration. Popularity produces premises obliteration, not because premises are processed almost unconsciously, as for instance in the case of visual perception, but because premises like symptomatic signs are not processed at all. This process obliterates all the symptomatic and relevant knowledge required for making a sound judgment. As if the adoption of a certain conclusion results from the bandwagon effect, then all the relevant and appropriate reasons for accepting it are excluded from the decision process.

Ultimately, the bandwagon effect deals with the adoption of a certain cultural variant regardless of its content, as the only reason which really counts is that *everybody is doing it*. In this sense, popularity is a very poor means of cultural transmission, since it just echoes an idea.[7] As Castro *et al.* [2004] argued, one of the key elements enabling eco-cognitive transmission is the capacity to approve and disapprove. This capacity underlies any process of learning, making imitation more reliable and accurate, because imitation is somehow driven by categorizing "one's own and others' behavior in terms of values – positive or negative, good or bad" [Castro *et al.*, 2004, p. 727]. This capacity, for instance, helps people avoid the costs related to trial-and-error learning. More generally, it is fundamental in causing some ideas to persist instead of others, and thus transforming an *imitation* system into a

---

[6] Magnani [2009] describes this process as underlying visual perception and other related forms. See also the semi-encapsulated nature of affordance detection that I will describe in section 4.4.

[7] Sunstein [2005b] developed this idea referring to the notion of "echo chamber". He brilliantly argued that echo chambers do not represent any model of information transmission or aggregation, but they simply lead to wild errors, undue confidence, and group polarization.

cumulative one. Popularity is neither good or bad in itself, since it delivers a way of assessing ideas or behavior, which does not generate a system of evaluation based on competence, but it is based on exaggerating "existing biases in individual decision-making" [Efferson *et al.*, 2008, p. 57].

To sum up, the most negative aspects of the bandwagon effect are:

- it is an impediment to further development, impairing the creative process;
- the majority of people tend to direct their effort towards a single topic (popularity of the topic);
- dissent is stifled, and the cost associated with dissenting dramatically increases as the bandwagon effect grows in strength;
- conformity reduces the possibility of social learning making imitation less profitable;
- popular ideas are increasingly taken on face value and remain unproved by those new to them, since popularity replaces other forms of control and evaluation;
- as an idea increases in popularity, it becomes less well known.

This eco-cognitive impoverishment is clearly captured by the distinction between competence-dependent information and competence-independent information that I have already introduced. As we switch from the former to the latter, we immediately lose the chance to lean on resources that are based on knowledge and competence and thus more reliable to get the job done.

2.7 Concluding Remarks

In this chapter I have illustrated how the notion of biased rationality is a development of the idea of bounded rationality. I have presented its virtues but also its vices. I have briefly introduced the argument on de-biasing rationality. In the next chapter I will be dealing with the notion of the cognitive niche. I will show how cognitive niches play a crucial role in moving the bounds of rationality. Basically, I will point out that, even though we are still bounded, our cognition can be extended and, insofar as we construct ever more sophisticated cognitive niches, we have the chance to de-bias our rationality and cognition.

Chapter 3
Moving the Bonds: Distributing Cognition through Cognitive Niche Construction

Introduction

In the previous Chapter I discussed the virtues and vices of bounded rationality as biased rationality. Basically, I pointed out that biases and fallacies can be useful insofar as one lacks symptomatic information to make a decision. In this sense, rationality, and human cognition, is always bounded, meaning that we are always committed to cognitive economy. However, rationality and our cognition can also be *de-biased*. On this point, I have argued that one way of de-biasing human cognition is to extend it. What does this mean? This means that humans improve the quality of their decisions and results by means of building up eco-cognitive structures that deliver more symptomatic information. The present chapter is mainly devoted to exploring in detail the distributed dimension of cognition. Human beings overcome their limitations by distributing and securing cognitive functions within their environment. This process does not lead to a complete de-bounding of the human cognitive system, rather, it contributes to pushing the bounds of cognition.

In section 3.1, I shall set the scene for the rest of the chapter illustrating the cognitive relevance of what it is called *the externalization process*. Basically, human beings overcome their internal limitations by (1) *disembodying* thoughts and then (2) *re-projecting* internally that occurring outside to find new ways of thinking. Accounts of distributed cognition seem to have overlooked this process placing more attention upon the first half. Conversely, in this section I shall see how it is the interplay between the organism and its environment that is responsible for moving the bounds of cognition.

This will allow us to introduce the concept of cognition as a *chance-seeking system*. Chance-seeking is an important part of de-biasing rationality as in extending their cognition human beings do not actually hold a complete representation of their environment, but they simply make use of anchors, which are literally *picked up* upon occasion for solving problems. Thus, in this respect, human cognitive behavior consists in *acting upon* those anchors which we ourselves have secured a cognitive function to.

E. Bardone: Seeking Chances, COSMOS 13, pp. 47–75, 2011.
springerlink.com © Springer-Verlag Berlin Heidelberg 2011

The idea of humans as agents seeking chances will be developed in section 3.2. In that section, I will introduce the idea of distributed cognition and I will illustrate the extended model of rationality. The model will be presented by comparing it with the other models of rationality treated in Chapter 2.

The activity of *chance seeking* is developed in sections 3.3 and 3.4 within an evolutionary framework based on *niche construction*. In my view, the theory of niche construction is the best theory for dealing with the evolutionary aspects of knowledge. Extending this theory from the biological domain to the cognitive one, I will show how the cognitive asset enhancement I talked about in section 2.5.1 is made possible at the *eco-cognitive level*. Basically, a *cognitive niche* provides humans with an additional source of information storage and computational abilities, which support and even boost the capacity of exhibiting ever more flexible adaptive responses to an ever-changing environment. These extra-genetic materials, properly exploited by ontogenetic mechanisms like learning, provide the unique framework for re-adjusting and refining our cognitive assets.

In section 3.5, I will illustrate how the approach based on niche construction can fruitfully account for some phenomena related to the so-called group selection. My take will be that groups enter the evolutionary scene as far as they allow cognitive niches to persist. As to this issue, I will introduce the notion of cognitive niche maintenance as underlying some of the activities, which are responsible for the persistence of ecologically delivered non-genetic information.

3.1 Humans as Chance Seekers

3.1.1 Incomplete Information and Chance-Seeking

Humans usually make decisions and solve problems relying on incomplete information [Simon, 1955]. Having incomplete information means that 1) our deliberations and decisions are never *the best* possible answer, but they are at least *satisficing*; 2) our conclusions are always *withdrawable* (i.e. questionable, or never final). That is, once we get more information about a certain situation we can always revise our previous decisions and think of alternative pathways that we could not "see" before; 3) a great part of our job is devoted to elaborating conjectures or hypotheses in order to obtain more adequate information. Making conjectures is essentially an act that in most cases consists in manipulating our problem, and the representation we have of it, so that we may eventually acquire/create more "valuable" knowledge resources. Conjectures can be either the fruit of an abductive selection in a set of pre-stored hypotheses or the creation of new ones, like in scientific discovery (see section 4.3). To make conjectures humans often need more evidence/data: in many cases this further cognitive action is the only way to simply make possible (or at least enhance) a thought to "hypotheses" which are hard to successfully produce.

Consider, for instance, diagnostic settings: often the information available does not allow a physician to make a precise diagnosis. Therefore, he/she has to perform additional tests, or even try some different treatments to uncover symptoms otherwise hidden. In doing so he/she is simply aiming at increasing the *chances* of making the appropriate decision. There are plenty of situations of that kind. For example, scientists are continuously engaged in a process of manipulating their research settings in order to get more valuable information, as illustrated by Magnani [2001]. Most of this work is completely tacit and embodied in practice. The role of various laboratory artifacts is a clear example, but also in everyday life people daily face complex situations which require knowledge and manipulative expertise of various kinds no matter who they are, whether teachers, policy makers, politicians, judges, workers, students, or simply wives, husbands, friends, sons, daughters, and so on. In this sense, humans can be considered *chance seekers*, because they are continuously engaged in a process of building up and then extracting latent possibilities to uncover new valuable information and knowledge.

The line of thought I will try to develop in the course of this chapter is the following: as chance seekers, humans are *ecological engineers*. That is: humans like other creatures do not simply live their environment, but they actively shape and change it looking for suitable chances. In doing so, they construct *cognitive niches* [Tooby and DeVore, 1987; Pinker, 2003] through which the offerings provided by the environment in terms of cognitive possibilities are appropriately selected and/or manufactured to enhance their fitness as chance seekers. Hence, this ecological approach aims at understanding cognitive systems in terms of their *environmental situatedness* [Clancey, 1997; Magnani, 2005]. Within this framework, a chance is that "information" which is not stored internally in memory or already available in an external reserve but that has to be "extracted" and then *picked up* upon occasion.

Related to this perspective is also the so-called *Perceptual Activity Theory* (PA) [Ellis, 1995; Ramachandran and Hirstein, 1997]. What these studies suggest is that an observer actively selects the perceptual information it needs to control its behavior in the world [Thomas, 1999]. In this sense, we do not store descriptions of pictures, objects or scenes we perceive in a static way: we continuously adjust and refine our perspective through further *perceptual exploration* that allows us to get a more detailed understanding. As Thomas [1999] put it, "PA theory, like *active vision* robotics, views it [perception] as a continual process of active interrogation of the environment". As I will show in the following sections "the active interrogation of the environment" is also at the root of the evolution of our organism and its cognitive system.

3.1.2 The Externalization Process

In the previous chapter, I claimed that human cognitive performances cannot be regarded as activities brought about solely by the isolated brain. In contrast, as already mentioned, humans lean on external objects. However, it is worth noting that this dependence is not a passive process, but it is *active*. In other terms, cognitive systems are not affected by external "inputs" that set the right configuration of our brain. Look, for example, at the case of language. Language, as an artifact, allows people to capture events in words [Donald, 2001; Harris, 2004; Love, 2004; Menary, 2007]. Consider, for instance, what our consciousness or our thought capacity would be without it. As a matter of fact, talking about things or, better, writing about them dramatically improves the quality of our thoughts. I argue that the action of writing allows humans to reach at least two advancements: (a) people can externally *reproduce* something that they have only within the *isolated* brain and make it more visible; (b) once they have externalized their thoughts in external objects, people can work on them and develop *new* concepts and new ways of thinking. I call the entire cycle the "externalization process," and it can be summarized as follows: human beings overcome their internal limitations by (1) externalizing and disembodying thoughts, ideas, solutions, and then (2) re-projecting internally that occurring outside in the external invented structure to find new ways of thinking.

It is worth noting that during the externalization process individuals create something that exists without their brain too; that renders the structure invented (an artifact or a tool) eventually useful to other individuals. Therefore, the externalization process constitutes the basis for social interactions.

As shown above, the externalization process allows us to make thoughts, ideas, etc. more visible. I call this process *mimetic*, because individuals use external supports to *reproduce* what occurs inside their private consciousness [Magnani, 2006b]. This is a very common experience for human beings. For example, writing is a mimetic activity, because we *represent* and *reproduce* thoughts, ideas, etc., in another means (the sheet of paper). Besides, the process of externalization makes it possible to use external objects, and the environment in general, as *information storages* that people can take advantage of in many ways [Mithen, 1996; Mithen, 1999; Donald, 2001]. Therefore, the activities of externalizing are mimetic, because humans try to reproduce (private) thoughts, feelings, ideas, and so on, in external objects that become information stores.

The example of writing is interesting in another respect to that I mentioned above. Once our thoughts have been secured to an external support (the sheet of paper), we are able to think and operate on them in a way that would not otherwise be possible. As a matter of fact, we cannot *re-read* our thoughts, because they are fleeting and immediately fade away. But, once written, we can use the sheet of paper as a *cognitive and epistemic mediator* [Hutchins, 1995; Magnani, 2001] and perform some cognitive activities otherwise impossible. More precisely, external supports allow individuals to *re-project* their own thoughts so that they can uncover hidden information and concepts. In this sense, external objects do not simply help to accomplish some cognitive activities – serving as cognitive mediators, but they allow us to find room for new ones. Thus, the activities of re-projecting are creative: external supports function as cognitive and epistemic mediators that find room for concepts and new ways of inferring which cannot be found internally ("in the mind").

Both mimetic and re-projecting activities organize human brains: more precisely, they make new configurations of neural networks and chemical process possible [Magnani, 2006a]. In neurological terms, novel experiences strengthen some synapses, weaken others, and they create new pathways amongst some sets of interneurons [Rose, 2005, p.160].

3.2 Bounds Moved: From Bounded to Distributed Cognition

The line of thought I am trying to pursue, leads us to point to a few relevant features of the original model of bounded rationality I have described in Chapter 2. In synthesis, the major claims are that bounded rationality

1. relies on computational capabilities only,
2. establishes a computational procedure to reach satisficing results,
3. doesn't recognize the role of internal and external variables not directly connected to computation.

These points make the model out to be a static one. In order to avoid these problems in understanding and describing human behavior, I present a model of rationality based on recent theories of distributed cognition. The starting point is the recognition of the role played by internal and external resources in human cognition.

3.2.1 Internal and External Resources

Some of the critiques on the original bounded rationality model refer to the dichotomy between environmental and internal resources. The two spheres can be thought of as internal and external constraints to bounded rationality [Todd and Gigerenzer, 2003]. They insist on the fact that "there is another possibility regarding the bounds, external and internal, that surround our rationality: rather than being separate and unrelated, the two sets of bounds may be intimately linked" [Todd and Gigerenzer, 2003, p. 144].

The merger confers dynamism to the original BRM, as "the internal bounds comprising the capacities of the cognitive system can be shaped, for instance by evolution or development, to take advantage of the structure of the external environment" [Todd and Gigerenzer, 2003, p. 144]. That is to say that the two "bounds" are strictly interconnected, as the external ones modify (or "shape", as they put it) the internal ones. Thus, bounded rationality is the "positive outcome of the two types of bounds fitting together. In other words, humans exhibit ecological rationality, making good decisions with mental mechanisms whose internal structure can exploit the external information structures available in the environment" [Todd and Gigerenzer, 2003, p. 144].

Todd and Gigerenzer's intuition is very original and is developed through a so-called "ecological rationality research program." It consists in defining heuristics – i.e. the way individuals gather and process data related to a specific problem – that match specific scenarios [Todd and Gigerenzer, 2003, p. 148]. Selected heuristics are very simple, and the Authors' thesis shows how effective behavior can also be explained by "fast and frugal" mechanisms. Following this framework, they find that "there are cases where cognitive limitations actually seem to be beneficial, enabling new functions that would be absent without them, rather than constraining possible behaviors of the system" [Todd and Gigerenzer, 2003, p. 160].

The merger between the external and internal resources leads to a new model where human bounded rationality "filters" the external variables and shapes its boundaries. However, this fundamental aspect is not integrated in the research program where schemes of heuristics remain fixed, and rationality concerns the choosing of a preferred scheme in relation to a given environmental context. If so, where is cognitive re-shaping located? And, what kind of restructuring are we facing? Moreover, what is the real impact of external resources on rationality? If external resources re-shape the cognitive system, can we think of it as mechanism where bounds are constantly moving?

3.2.2   The Role of External Representations

Recent research in cognitive science underlines the role of external resources in understanding how human cognition works.[1]

As a matter of fact, people constantly and heavily lean on external supports, and the quality of their performance would immediately drop down without them [Clark and Chalmers, 1998]. Humans constantly delegate cognitive functions to the environment: remembering and calculating, for instance, are heavily supported by the environment [Norman, 1999b]. Very simple artifacts, such as pen and paper allow us to accomplish tasks that otherwise we couldn't even think about [Donald, 2001]. And so forth.

All these external objects are not mere "approximating mechanisms" or supports, but they play a crucial role in *extending* the rationality of human behavior and decision-making. Bounded rationality theory, and its groundwork, fails to recognize the cognitive role exhibited by external objects. More precisely, since bounded rationality focuses only upon what goes on within the individual mind, it fails to account for the fact that external computational resources extend the rational capabilities of humans. Thus, I may claim that rationality is *un-bounded* from the confines of the limited individual brain by the exploitation of external resources.

In order to make this point, I shall deal with the concept of external representations. More precisely, we need to show how external resources play a crucial cognitive role in extending the rational character of human decision-making; they encode computational resources that can be fruitfully exploited by humans to overcome their cognitive limits.

Generally speaking, a problem can be defined through an "initial state", a "goal state", and a "set of operators" (or mediators) that allow transformation of the initial state into the goal state by a series of intermediate steps. It is worth noting that the standard approach to decision making is not far from this representation. The

---

[1] For a general account of the distributed cognition approach, see [Norman, 1993; Salomon, 1993; Hutchins, 1995; Clark, 1997; Kirsh, 1999; Donald, 2001; Wilson, 2004; Clark, 2008; Rupert, 2010]. On the role of distributed cognition in science, see the concept of construal [Gooding, 1994], and that of epistemic mediator [Magnani, 2001]. For a general account of the moral role played by external resources, see the concept of moral mediator [Magnani, 2007c]. For a distributed cognition approach on the interaction between humans and computers (HCI), see [Norman, 1999a; Hollan *et al.*, 2000; Susi and Ziemke, 2001; Calvi and Magnani, 2002; Kirsh, 2004; Perry, 2003; Magnani and Bardone, 2006].

intermediate steps, that I will call hereafter "actions", can be grouped into two main categories: pragmatic and epistemic [Kirsh and Maglio, 1994]. By the term *pragmatic actions* I refer to all those intermediate steps that alter the world to achieve some physical goal or other physical intermediate stages. For example, if one has to be refunded for a certain purchase, he has to fax the receipt. The action of faxing the document is a pragmatic action because it brings one closer to the goal state, namely, being refunded. In contrast, "epistemic actions" are all those actions that alter the representation of the task one is facing. A child that shakes and manipulates their birthday present to guess what there is inside is a fair example of this kind; the action of shaking unearths additional information that makes guessing less blind [Magnani, 2001]. In this case, the world is not strictly changed: what is changed is the representation we have about the problem. Accordingly, epistemic actions can also be regarded as "task-transforming representations" [Hutchins, 1995].

What suggested above points to the conclusion that solving a problem means representing it so as to make the solution transparent [Simon, 2005]. Hence, the question is: how can we make the solution of a problem more transparent? What can make the solution more transparent? I claim that the cognitive role of external resources is precisely connected with shaping the representation of a task so as to transform difficult tasks into ones that can be easily carried out. Let me make an example. Consider, for instance, the following two medical prescriptions [Norman, 1993]:

Inderal	1 tablet 3 times a day
Lanoxin	1 tablet every a.m.
Carafate	1 tablet before meals and at bedtime

	Br	L	D	Bt
Inderal	x	x	x	
Quinaglute	x	x	x	x
Lanoxin	x			

Now, suppose we should answer the question "how many pills should I take at lunch time?" Here we have two different ways of representing the problem. The first is a traditional medical prescription that simply tells us what kind of pills we should take, whereas the second is a matrix. If we consider the two representations we immediately come up with the conclusion that the way the second represents the task is much easier than the first. The reason being the one suggested by Simon, that is, that matrix representation makes the solution more transparent. The medical prescription in the table above is far more complex. Already in the first line we need to think about what "1 tablet 3 times a day" means. Once we came up with the number of pills we should take, we have to write it down. Then pass to the second line, and so forth. In contrast, the second representation is much simpler: answering the question simply means scanning down the lunch column, and counting the colored squares. We may even say that one gets the answer at a glance.

Following these hypotheses, numbers of scholars [Zhang, 1997; Gatti and Magnani, 2006; Knuuttila and Honkela, 2005] argue that the tradi-tional notion of representation as a kind of abstract mental structure is misleading. As the example shows, some cognitive performances can be viewed as the result of smart interplay between humans and the environment. The figure below illustrates our point.

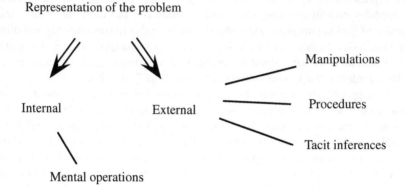

The representation of the task we face is only partly internal. That is, when we try to accomplish a certain task, we exploit computational and cognitive resources embodied into external objects: for we are often engaged in such processes without holding an explicit and internal representation of them. In this case, an external representation is involved in terms of the actions, procedures, and tacit inferences we are actually triggered to carry out. More generally, I may argue that external representations can be considered as "tacit procedures" [Polanyi, 1966] that emerge from, and are prompted by, the interaction between humans and the environment. Therefore, internal representation does not *mirror* the entire representational task, because it is only a part of it.

3.2.3   Broad Cognitive Systems

In the last paragraph I pointed out how humans constantly lean on external resources to accomplish various tasks. I have outlined my approach relying on the notion of external representation. In my view, this notion plays a key role in understanding how external objects and symbols can enhance human capabilities. In this paragraph I shall detail some consequences that this approach brings about dealing with the notion of the cognitive system.

That the environment plays an active role in shaping decision-making activities is based on the assumption that a cognitive system goes beyond the confines of the skull [Clark, 2003]. That is, the skull is not a "magic" boundary that clearly distinguishes what counts as cognitive and what does not [Wilson, 2004]. There are several activities and performances that cannot be carried out only by the naked brain. External resources actively shape cognitive performances that cannot be "tax-onomized individualistically", say, only referring to what happens within the brain [Wilson, 103, p. 352]. Some cognitive processes that we attribute only to humans are the result of smart interplay between humans and the environment. According to that, cognitive systems can be viewed as a set of "packages of resources and operations" [Clark and Chalmers, 1998, p. 14]. This set is open to external upgrades and changes, and most of all is distributed. Indeed, the brain operates on a package of basic cognitive resources, but the reason why we praise it so much is because of its "portability" [Clark and Chalmers, 1998].

This conceptual branching leads to two main points. First of all, external resources can support pre-existing abilities such as memorize or remembering. External symbols, that include rudimentary technologies, release humans from the limitations of the brain's biological memory systems [Donald, 2001].

Secondly, external objects can also bring into existence additional cognitive abilities that the naked brain could not exhibit by itself. For example, there are several instances pointing to the conclusion that anthropomorphic thinking was brought about through the mediation of external objects that made it possible to integrate the two separate intelligence [Mithen, 1999; Magnani, 2006a]. Upper Paleolithic cave paintings seemed to be fundamental aids to our ancestors in order to store information about animal location and behavior [Eashtem and Easthem, 1991; D'Errico and Cacho, 1994; Mithen, 1996; Lewis-Williams, 2002]. They were supposed to be models or maps for the specific terrain around the caves so that predictions about the natural world were improved and decision-making facilitated.

The external resources approach broadens and deepens the original concept of rationality. Rationality is not referred to the decision-making process here while the result one obtains through the delegation of cognition to definite external resources is rational. It clearly appears then, that we do not refer merely to the computational capabilities but, more extensively, to the way human cognition is shaped and extended when getting in contact with external resources. Stating "when getting in contact to external resources" means in every circumstance, always; though, rationality is "expanded" by the way it depends on external resources and is enhanced by them.

Let me make another example following on from Simon. He recognized that "[t]he introduction of computers changed the ways in which executives were able to reach decisions; they could now view them in terms of a much wider set of inter-related consequences than before. The perception of the environment of a decision is a function of – among other things – the information sources and computational capabilities of the executives who make it" [Simon, 1978, p. 8]. Here we have two questions. Firstly, can we suppose that the executive changes his attitudes towards problem solving and decision making when the computer is introduced? Secondly, has the user enhanced her/his computational abilities in problem solving and decision making?

I may say that Simon should have positively answered the first question, but not the second one. According to bounded rationality theory, there is no account to explain whether human computational capabilities are developed and shaped through external resources or not.

The considerations made in the last two paragraphs might lead thinking that Simon's concept of rationality is a lighter version of what it actually is. Rationality as a computational resource can be defined as the under-esteemed version of the whole range of human rational capabilities. Thus, rationality is not bounded, while internal computational capabilities are. More precisely, computational capabilities (I) partially represent human decision-making processes, and (II) change – together with the other human attitudes (psychological, ethical, political, economical, etc.) – on their own and in relation to external resources. Following statement (I), I argue that rationality is not limited to computational capabilities, as we make decisions and obtain results also using not-entirely exploitable procedures. Then, following statement (II), I sustain that rationality is not bounded, neither in relation to the potential modifications of personal capabilities (think of the same individual as a child and as a Ph.D. laureate),[2] nor if we recognize the role of external resources (artifacts) in modifying our cognitive system. Broadly speaking, I refer to the role played by (1) time and (2) representations.

In summary, indeed, we do have limitations, but these are always changing (the cognitive system is not stable by definition) and heavily dependent on external resources. In this way, it is clear that our model cannot be confused with the unbounded rationality model of the neoclassical approach.

3.2.4   The Extended Model

In order to reach a better understanding of the differences between the four models and to get a clearer picture of our model-building process (see Table 3.1). Here I try to express the main differences between the four models – neoclassical (NCM), bounded (BRM), biased (BiRM) and the extended rationality model (ERM). Bounded rationality was our starting point while the neoclassical approach remained in the background. Nevertheless, it is very interesting to show differences and similarities between the four as, in my opinion, bounded and biased rationality maintain significant links to the traditional economic model especially with respect to the distributed nature of cognition.

I define and compare the three models using five variables:

1. the kind of result (or solution) attained by each model;
2. the procedure leading to that result;
3. the hypotheses the models put forward on the human cognitive system;
4. the cognitive capabilities one is supposed to have; and,
5. the philosophical meaning attributed (mostly implicitly) to the acting entity, i.e. the carrier.

---

[2] It is not necessarily true that a Ph.D. laureate develops superior capabilities if compared to him/herself as a child; we are only trying to differentiate.

The variables are not sorted by their importance. But, they can be organized thinking of the philosophical hypotheses on the individual (carrier) first, and on the other variables as a subsequent implication of that assumption. For example, the carrier has definite cognitive capabilities that are related to her/his cognitive system which uses a particular procedure in order to obtain the result. For a better explanation of the model's workings, it should be clearest to start from the first column. The major part of the issues here cited, are defined and criticized above.

Table 3.1 Models of Rationality

	Cognitive system	Cognitive capabilities	Carrier	Result	Procedure
Neo-classical model	Brain-in-its-box	Unbounded	God-like creatures	Optimal	Brute force
Bounded rationality	Brain-in-its-box	Bounded	Humans	Sub-optimal	Brute force
Biased rationality	Brain-in-its-box	Bounded	Humans	Sub-optimal	Bias&Fallacy
Extended rationality	Distributed	Extended	Humans	Workable	Externalization

The four models define in various ways the results their decision-maker obtains. The neoclassical model refers to optimal results, i.e. the best, while through the bounded and biased rationality model we obtain only sub-optimal results. The extended model suggests that the individual gets results that usually fit a particular situation, and are not necessarily linked to the optimum result. They are workable in the sense that they allow us to manage definite situations and can be modified (or improved) when conditions change, both internal and external ones. This is in line with Beach and Mitchell's image theory [Beach, 1998]; the cognitive model I provide can serve as a useful basis for their approach.

My model is based on the assumption that both the NCM and the BRM use computation (or brute force strategy) as the only rational goal-attainment procedure.[3] In BiRM and ERM the procedure depends on the joint activity of internal and external resources, so that computation is only one possible procedure. Emotions, morality (in terms of personal values), ideas of justice and fairness, culture, etc. have to be integrated into the rational model of choice between alternatives. in ERM, however, the meaning here attributed to the exploitation of external resources is strictly linked to the cognitive meaning attributed. In other words, if all these elements occur in the decision-making process, it is clear that computation is a specific case, that cannot be considered in isolation. It can be prevalent or not, but it does not work as the three other models suppose it does.

The three models all suppose that the cognitive system is not distributed. In the first case (NCM), the system works in complete isolation; in the second case (BRM) the environment is a source of constraint to the "natural" and "static" human brain.

---

[3] This point is developed in quite detail by Patokorpi [2008].

In the case of BiRM, contextual variables are taken into account, but this ecological element is not integrated into a cognitive framework therefore stressing the distributed nature of cognition. The ERM, on the contrary, is based on the distributed cognitive system, in the sense that external resources define our system and the way it works.

The main point here is that of limits to individuals' cognitive capabilities. The NCM hypothesis is that of individuals having no limits; we say they are unbounded [Shakun, 2001]. On the contrary, BRM and BiRM computational capabilities, as the only procedure to obtain results, are limited in individuals. In my approach, these rational bounds do exist as observed in actual human behavior; however, they are related (a) to the use and meanings of external resources, (b) to the general social environment in which the individual behaves, and (c) to the time-effect (also reflected on means-end modifications). The individual here considered is clearly a god-like creature, for the NCM, while it is human for the other two. It is clear if we consider the normative flavor of the first and the behavioral intent of the other three.

The bounded rationality model and the biased rationality model defines rationality as internal computational capabilities, emphasizing the use of brute force. As a consequence, it cannot explain a series of successful results that do not depend on computation. I linked these missing points to the fact that the model overlooks important variables such as the role of representations and external resources affecting the human cognitive system.

As I have tried to outline, recent cognitive studies highlight that the way our mind operates are very different from what is supposed by the BR model and the BiR model. Instead of definition based on negative assumptions – neoclassical denial – we have to find the positive attributes of human rationality, and to redefine it on the basis of the human cognition system.[4] Assuming that the human cognitive system is shaped by external resources and representations, implies a few fundamental points:

1. decision making activities (and the way to obtain successful results) derive from the way individuals interact with the environment;
2. this interaction involves internal and external resources, and the way they are represented;
3. this process is dynamic, in the sense that cognitive capabilities depend on the exploitation of external resources and on their representation (time and way of modifying the interaction);
4. interaction and dynamics imply uncertainty and complexity, in terms of difficulties deciphering between internal and external influences;
5. the "smart interplay" between the two is not limited to computational capabilities and, even if we narrowly focus on them, they are not only internal, but depend from the "smart interplay" itself;
6. thus, our computational capabilities are not limited, since bounds depend on the "smart interplay" between internal and external resources and, moreover, the result of the decision making process is embedded in the way the broad cognitive system employs, represents, and acknowledges external resources.

---

[4] This is the line of thought developed, for instance, by Secchi [2010].

These six points suggest thinking about rationality and decision making on the basis of a different approach that enriches the BR model. In particular, the partial results of this contribution can be highlighted as follows:

1. we are able to define a cognitive model that connects rationality to the way individuals employ their cognitive system;
2. if the role of external resources and representations is that defined above, this approach questions the neoclassical, the BR, and BiR models, concerning its basic assumptions on computational capabilities;
3. we call for a new model of rational choice, that needs to be effectively based on external resources and representations interplay, in order to take rationality's "moving bounds" into account;
4. the model based on distributed cognition finds an easier way to explain the role of critical issues, such as emotions, since they are treated as representations.

From these starting points, the analysis of the model of rationality can follow a number of different tracks. However, I am focusing on three complementary issues. The first one is related to the decision making process in general terms, where I try to discover the main underlying variables of human rationality.

The second issue is devoted to the analysis of decision making in an organizational context. This is the line I am following in order to define a new model of rationality based on the distributed cognition approach, as organizations normally provide researchers with more limited contexts where motivation, creativity, leadership, intuition, staff-line relationships, etc. are easier to recognize and to study.

The last but not least, an important issue of our research activity is empirical testing of the model. This part is not clearly separated from the other two, but is fundamental for the trial and error process of outlining a new model. Both the study of decision-making in general and as it is applied to organizations need to be supported by data, in order to reach a clearer and more useful model or to switch to another one. The main variables we focus on are external resources, representations, and, in general terms, the environment in which decisions are made.

I started from the assumption that the modern social sciences, and especially economics and management, need to be strongly rooted in actual behavior. The model of rationality based on the distributed cognition approach is basically an attempt to go further this direction: analyzing human rationality as it actually is.

3.3 Cognitive Niche Construction: Distributed Cognition Evolving

In this section, I am going to shed light on the evolutionary dimension that the distributed cognition approach developed so far may have. In doing so, I will rely on the notion of niche construction – a topic that has been neglected for a long time in biological studies. Niche construction theory will provide a suitable framework for discussing some fundamental issues related to the idea of cognition discussed in the previous sections.

3.3.1 Niche Construction: The Neglected Side of Evolution

It is well-known that one of the main forces that shape the process of adaptation is natural selection. That is, the evolution of organisms can be viewed as the result of a selective pressure that renders them well-suited to their environments. Adaptation is therefore considered as a sort of *top-down process* that goes from the environment to the living creature [Godfrey-Smith, 1998]. In contrast to that, a small fraction of evolutionary biologists have recently tried to provide an alternative theoretical framework by emphasizing the role of niche construction [Laland *et al.*, 2000; Laland *et al.*, 2001; Odling-Smee *et al.*, 2003].

According to this view, the environment is a sort of "global market" that provides living creatures with unlimited possibilities. Indeed, not all the possibilities that the environment offers can be exploited by the human and non-human animals that act on it. For instance, the environment provides organisms with water to swim in, air to fly in, flat surfaces to walk on, and so on. However, no creatures are fully able to take advantage of all of them. Moreover, all organisms try to modify their surroundings in order to better exploit those elements that suit them and eliminate or mitigate the effect of the negative ones.

This process of *environmental selection* [Odling-Smee, 1988] allows living creatures to build and shape the "ecological niches". An ecological niche can be defined, following Gibson, as a "setting of environmental features that are suitable for an animal" [Gibson, 1979]. It differs from the notion of habitat in the sense that the niche describes *how* an organism lives its environment, whereas habitat simply describes *where* an organism lives.

In any ecological niche, the selective pressure of the *local* environment is drastically modified by organisms in order to lessen the negative impacts of all those elements which they are not suited to. This new perspective constitutes a radical departure from traditional theory of evolution introducing a second inheritance system called *ecological inheritance system* [Odling-Smee *et al.*, 2003]. According to this view, acquired characters – discarded for such a long time – can enter evolutionary theories as far as they cause a modification to the environment that can persist and thus can modify the local selective pressure.[5] Ecological inheritance system is different from the genetic one in the following way [Odling-Smee *et al.*, 2003]:

1. genetic materials can be inherited only from parents or relatives. Conversely, modifications on the environment can affect everyone, no matter who he/she is. It may regard unrelated organisms also belonging to other species. There are several global phenomena such as climate change that regard human beings, but also the entire ecosystem;

---

[5] This perspective has generated some controversies, since it is not clear the extent to which modifications count as niche-construction, and so enter the evolutionary scene. The main objection regards how far individual or even collective actions can really have ecological effects, whether they are integrated or merely aggregated changes. On this point, see Sterelny [2005] and the more critical view held by Dawkins [2004]. For a reply to these objections, see Laland *et al.* [2005].

2. genes transmission is a one way transmission flow, from parents to offspring, whereas environmental information can travel backward affecting several generations. Pollution, for instance, affects young as well as old people;
3. genetic inheritance can happen once during one's life, at the time of reproductive phase. In contrast, ecological information can be transferred during the entire duration of life. Indeed, it depends on the eco-engineering capacities at play;
4. genetic inheritance system leans on the presence of *replicators*, whereas the ecological inheritance system leans on the *persistence* of whatsoever changes made upon the environment.

Indeed, natural selection is somehow not halted by niche construction. Rather, this means adaptation cannot be only considered by referring to the agency of the environment, but also to that of the organism acting on it. In this sense, organisms are ecological engineers, because they do not simply live their environment, but they actively shape and change it [Odling-Smee *et al.*, 2003].

3.3.2 The Notion of Cognitive Niche

My contention is that the notion of niche construction can be also usefully applied to human cognition. More precisely, I claim that cognitive niche construction can be considered as one of the most distinctive traits of human cognition.

Organisms are equipped with various ontogenetic mechanisms that permit them to acquire information and thus better adapt to the environment: for instance, immune system in vertebrates and brain-based learning in animals and humans. Their role is to provide organisms with a supplementary mechanism to acquire information and thus face various environmental contingencies that are not – and cannot be – specified at the genetic level [Odling-Smee *et al.*, 2003, p. 255]. A genetically specified initial set of behaviors is elaborated through experience of a relevant environment. These ontogenetic mechanisms are therefore a sort of *on-board* system allowing flexibility and plasticity of response to an ever-changing environment, which are at the core of the notion of cognition I endorse.[6]

In the case of human beings and other mammals, bigger brains allow to store information which could not be pre-defined by the genes [Aunger, 2002, pp. 182–193]. Flexibility and plasticity of response to an ever-changing environment are connected to the necessity of having other means for acquiring information, more readily and quickly of the genetic one. I posit that niche construction plays a fundamental role to meet this requirement. Plasticity and flexibility depend on niche construction as far as various organisms may alter local selective pressure via niche construction, and thus increase their chances for surviving. More specifically, cognitive niches are crucial in developing more and more sophisticated forms of

---

[6] Godfrey-Smith [2002] defined cognition as the capacity of coping with a range of possible behavioral options with different consequences for the organism's chance to survive. This definition allows him to embrace a broader notion of cognition which extends it to animal and plant behaviors. I will embrace this thesis in the sections devoted to abduction and affordance.

flexibility, because they constitute an additional source of information favoring behavior and development control. In this case, epigenesis is therefore augmented, and, at genetic level, it is favored by genes regulating *epigenetic openness* [Sinha, 2006]. Epigenetic openness is closely related to what Godfrey-Smith [2002] called *phenotypic plasticity*; the flexible response of living organisms (humans in particular) leans on sensitivity to environment clues, and this process of attunement to relevant aspects of the environment cannot be separated from niche construction.

In the case of human beings, the process of attunement leans on the continuous interplay between individuals and the environment, in which they more or less tacitly manipulate what is occurring outside at the level of the various structures of the environment in a way that is suited to them. It emerges from a network of continuous interplay between individuals and the environment, in which they more or less tacitly manipulate what is occurring outside at the level of the various structures of the environment in a way that is suited to them. Accordingly, I may argue that the creation of cognitive niches is *the* way cognition evolves, and humans can be considered as ecological cognitive engineers.

3.3.3 Cognitive Niches and Distributed Cognition

Recent studies on distributed cognition seem to support our claim.[7] As already mentioned in section 3.2, problem solving or decision-making, cannot only be regarded as internal processes that occur within the isolated brain. Through the process of niche creation humans extend their minds into the material world, exploiting various external resources. Therefore, they exhibit a range of cognitive behaviors insofar as they are merged into a network of ecological interactions. For "external resources" I mean everything that is not inside the human brain, and that could be of some help in the process of deciding, thinking about, or using something. Therefore, external resources can be artifacts, tools, objects, and so on. Problem solving, such as general decision-making activity [Bardone and Secchi, 2009], for example, are unthinkable without the process of connection between internal and external resources.

In other words, the exploitation of external resources is the process which allows the human cognitive system to be shaped by environmental (or contingency) elements. According to this statement, I may argue that external resources play a pivotal role in almost any cognitive process. Something important must still be added, and it deals with the notion of representation: in this perspective the traditional notion of representation as a kind of abstract mental structure becomes old-fashioned and misleading.[8] If some cognitive performances can be viewed as the result of a smart interplay between humans and the environment, the representation of a problem is partly internal but it also depends on the smart interplay between the individual and the environment.

---

[7] Cf. Zhang [1997], Hutchins [1995], Clark and Chalmers [1998], [Wilson, 2004], Magnani; Magnani [2006a; 2007c].

[8] Cf. Zhang [1997], Gatti and Magnani [2006], Knuuttila and Honkela [2005].

An alternative definition of the ecological niche that I find appealing in treating our problem has been provided by Gibson [1979]: he pointed out that a niche can be seen as a set of *affordances*. My contention is that the notion of affordance may help provide sound answers to the various questions that come up with the problem of ecological niches. The notion of affordance is fundamental for two reasons. First of all, it defines the nature of the relationship between an agent and its environment, and the mutuality between them. Second, this notion may provide a general framework to illustrate humans as chance seekers. I will come back to this issue in the next chapter.

Within a distributed cognition framework, the activity of niche construction provides humans with the chance of externally storing and encoding a great deal of information and computational capabilities. Indeed, cognitive niches contribute to release a large amount of resources, for instance, in terms of knowledge storage and computational capabilities to manipulate information. Here again the case of language is an example worth citing.[9]

That some kinds of cognitive processes originally designed for specific purposes turns out to be useful for others is fairly captured by the notion of *cognitive fluidity* [Mithen, 1996]. Basically, cognitive fluidity refers to the capacity of applying to heterogeneous domains forms of thinking originally designed for specific tasks.[10]. Mithen himself seemed to acknowledge the ecological and distributed dimension of cognitive fluidity. Cognitive fluidity relies on what can be thought primarily as a direct effect resulting from the (quit unique) human activity of eco-cognitive engineering.

In Mithen's own words:

> The clever trick that humans learnt was to disembody their minds into the material world around them: a linguistic utterance might be considered as a disembodied thought. But such utterances last just for a few seconds. Material culture endures [Mithen, 1999, p. 291].

The continuous manipulation of various external materialities contributes to uncover new chances, which can eventually allow to find room for new concepts [Magnani, 2006a]. Here again the case of anthropomorphic thinking is a valuable example. In Mithen's view, it was made available to the human mind literally by the

---

[9] This last contention is in line with what Logan argued on the evolution of human complexity [Logan, 2006, p. 150]. He argued that a new level of order emerges as a response to an *information overload*. Though speculative, his contention is worth quoting: "Writing and mathematical notation arose in Summer as a response to keeping track of the tributes farmers paid to the priests [...]. This gave rise to formal schools to teach the skills of reading [...], which in turn led to scholars and scholarship giving rise to its information overload, which in turn led to computers and computers gave rise to its information overload [...]".

[10] According to Mithen cognitive fluidity was the product of what he called a "cognitive big bang", which was accompanied by a radical change as the neurological re-organization of the brain occurred. That cognitive big bang produced remarkable human creative abilities, i.e., an almost limitless capacity of imagination, because knowledge and information could flow freely between behavioral domains.

exploitation of external configurations of signs, in which our ancestors were able to curve from mammoth ivory a half human/half animal figure. As he put it:

> An evolved mind is unlikely to have a natural home for this being, as such entities do not exist in the natural world, the mind needs new chances: so whereas evolved minds could think about humans by exploiting modules shaped by natural selection, and about lions by deploying content rich mental modules moulded by natural selection and about other lions by using other content rich modules from the natural history cognitive domain, how could one think about entities that were part human and part animal? Such entities had no home in the mind. [Mithen, 1999, p. 291]

External structures, which ultimately are meshed into our cognitive niches, exhibit what may be called a cognitive (semiotic) agency going beyond the individual. That is, once externalized and secured to external supports, ideas, thoughts, and even intentions, cease to be what they originally meant to be. They acquire a public status; that means they go under a process of negotiation, which eventually leads to conventionalization and/or entrenchment [Tylén, 2007]. This can be vieId also as an hybridization process, which not only regards human beings and their surroundings, but also those objects and artifacts that enter the cognitive niche. Secondly, human externalizations become part of the so-called eco-cognitive inheritance and, therefore, being subjected to further modifications and exploitations insofar as they can be also the basis for the creation and development of additional eco-cognitive capabilities.[11]

The neurological counterpart of this process is a process of brain re-configuration and re-organization – a rehearsed recapitulation – which allows our brain to disentangle itself from the perception-action cycle typical of the on-line thinking [Magnani, 2009].

3.4   The Future Enrichment of Cognitive Niches: The Case of Ambient Intelligence

The cognitive niches in which we live may indeed change. They should not be taken for granted. They may perish (cf. section 2.6) as well as progress. In this section I will be dealing with the latter case, the case in which our cognitive niches are *enriched*. In order to do so, I will discuss the case of Ambient Intelligence as cognitive niche enrichment. That is, I will contend that Ambient Intelligence strongly favors the development of some crucial and structural aspects of human cognition, in so far as it creates – given the tremendous technological advancements over recent decades – a new way in which humans can cognitively live their environments.

Ambient Intelligence surely represents a novelty with respect to how humans can re-distribute and manage the cognitive resources delegated to the environment. Although massive cognitive delegations have already been successfully completed,

---

[11] Some authors tried to model how ecologies – cognitive niches, in my terminology – would emerge and develop. For more information about this issue, see for instance Holt [2009] and for a more cognitive-oriented account of the same matter see Pata [2009].

Ambient Intelligence drastically favors the establishment of new kinds of environment which present novel theoretical and cognitive features, worthy of analysis.

I contend that, in analyzing the cognitive significance of Ambient Intelligence, the notion of *cognitive niche* is crucial. Such a notion would allow us to clear up possible misunderstandings about the role played by technological innovation in evolution. Besides, the notion of the cognitive niche will be of interest, as I will claim that Ambient Intelligence can be considered a new stage in the history of cognitive niche construction, given its "eco-cognitive" impact. Accordingly, I will claim *AmI* can be considered a form of *cognitive niche enrichment*.

The advantages of having such devices that mimic a human mind are not difficult to grasp. Basically, we are talking about the possibilities of having external artifacts able to *proactively* and sensitively assist people in a number of tasks. Ambient Intelligence adds a new layer to the traditional ways of disembodying the mind: Ambient Intelligence basically puts those sophisticated and smart devices – mimicking our mind – into our environments. In doing so even familiar objects may embed high-level computing power. More generally, I argue that Ambient intelligence deals not only with reproducing some kind of sophisticated human cognitive performance, but also focuses attention on the eco-cognitive dimension of computing – referred to as context-aware computing [Cook and Das, 2007].

Ambient Intelligence enriches the experience of our environment in many ways. The most striking aspect related to Ambient Intelligence as a form of distributive intelligence is the level of autonomy that smart environments can reach [Hildebrandt, 2008a]. A smart environment has an amazing power of monitoring and subsequently keeping track not only of our actions – what we do – but also of our preferences – what we desire [Remagnino *et al.*, 2005; Cook *et al.*, 2009]. Collecting such an amount of data – and aggregating it – allows smart environments to provide us with feedback that exhibit a degree of adaptability that cannot be compared with any other traditional environment (or cognitive niche, as I will show in the following sections). First of all, a smart environment adapts itself to infer one's preferences. It can act on the basis of past interactions that have been appropriately stored and then exploited by various tools mimicking some sophisticated forms of reasoning. This means that in smart environments high-level customization is possible, relying on the collection and aggregation of data about our behavior. These environments can also be creative insofar as they can anticipate user preferences, even before they become aware of them themselves [Hildebrandt, 2008a]. In this case, smart environments exhibit what Verbeek [2008] calls "composite intentionality". Basically, composite intentionality refers to situations in which the intentionality resulting from an action we take is made up of our own in coordination with that emerging from the interaction with an artefact. The intentionality resulting from interaction in smart environments is indeed highly composite, as *AmI* is designed specifically for augmenting – and thus making accessible – some experience of the world with respect to various modalities, namely, sensing, acting, and making decisions. I will come back to this issue in section 4.5 devoted to "adapting affordance".

Thus, to summarize, the idea behind distributed cognition is that human cognitive capabilities are fundamentally shaped by environmental chances that are ecologically rooted. Cognitive processes do not happen in a vacuum, so the context and the resources one has at one's disposal are crucial for describing and also explaining human cognition. Ambient Intelligence can certainly be considered one of the most sophisticated ways humans have invented to distribute cognitive functions to external objects. In this case, the massive cognitive delegation contributes to a radical re-distribution of the cognitive load humans are subjected to. Basically, Ambient Intelligence improves people's experience in their environments [Cook and Das, 2007]. That is, it increases the number and range of tasks one can accomplish. As the result of a massive cognitive delegation, humans are provided with environments that bring sophisticated interactions into existence, in which the cognitive load is partly carried out by intelligent devices displaying an unprecedented level of autonomy and transparency.

As already mentioned, the outstanding characteristics of human beings are that they have progressively become *eco-cognitively* dominant, as their abilities as eco-cognitive engineers out-competed the ones of other species in occupying and then modifying according to their needs the shared environments [Flinn *et al.*, 2005]. The notion of cognitive niche acknowledges the artificial nature of the environment humans live in, and their active part in shaping it with relation to their needs and goals. It is within this context that I locate the innovative character of Ambient Intelligence. That is, the creation of smart environments can certainly be viewed as a new way of constructing the cognitive niche. More precisely, Ambient Intelligence constitutes an eco-cognitive activity in which our pre-existing cognitive niches are dramatically enriched with objects and tools that re-configure our capacities for extending human cognition and its boundaries.

It is worth noting – even though I will not be dealing with this issue – that niche construction activities do not halt natural selection. This certainly has some consequences that should be acknowledged. Niche construction activities may help humans reduce the negative impacts of pre-existing niches but that does not mean that they might not produce even worse consequences. This is because, even when a niche has been modified, selection pressures continue to act upon us and other organisms as well. Certain cognitive niches resulting from intensive eco-cognitive activity may turn out to be "maladaptive" or – at least – they might actually endanger us and other species as well. Here, the term "maladaptive" is not intended in a Darwinian sense.

In Ambient Intelligence we have many examples of potential maladaptive outcomes resulting from cognitive niche enrichment activity. For example, the problem of the possible negative consequences of Ambient Intelligence in the case of agency and (criminal) liability is discussed in Hildebrandt [2008a]. Hildebrandt brilliantly points out that the emergence of ambient technologies that are able to monitor and anticipate human behavior can become a threat to a number of values that are crucial in Western democracies, like for instance, the values that are at play in the case of criminal liability. She argues that, as far as we cannot say whether a certain action has been carried out by us or by an artificial device, then we will also have severe problems in attributing criminal liability. This may also cause a person to exploit the ambiguity resulting from the *AmI* establishment of a hybrid agency, in their favor.[12] Another possible negative consequence is related to the notion of identity. As argued by Gudwirth [2009], *AmI* technologies empower humans in the process of profiling. Profiling basically deals with the possibility of assigning an identity to a user relying on data and information gathered from the behaviors of the same user as well as of others. The pervasiveness of *AmI* drastically changes our ability and the effectiveness of such a task, given the continuous and detailed monitoring of the user's behavior made possible by the smart devices available. *Being profiled*, as Gudwirth argued, could, however, easily become a threat to the development of our own identity because it can be assigned automatically, and even without our consent. In this sense, *AmI* potentially and dangerously induces us, first of all, to adopt an identity we did not have the opportunity to choose. This is clearly a limitation of our freedom insofar as we would be obliged – more or less tacitly – to match in with some arbitrary categories generated by the profiling algorithms. Secondly, it enforces us to adapt "to a context moulded by other actors" [Gutwirth, 2009] favoring various dynamics related to standardization and, sometimes, even group tyranny. [13]

In order to further and better assess the eco-cognitive contribution of Ambient Intelligence the notion of *affordance* is fundamental. In section 4.5 I will illustrate and develop the idea that the novelty of *AmI* lies in the fact that Ambient Intelligence enriches human cognitive niches by providing new affordances. I will show how affordances are cognitive chances embedded in the interaction between a (human) organism and its environment, and how they are organized in cognitive niches, which make them easily accessible. Ambient Intelligence populates our cognitive niches with objects and devices that are to some extent intelligent objects. More

---

[12] As for the relationship between modern law and technology, more generally, Hildebrandt [2008b] has recently argued that they are indeed coupled together, as technologies always exhibit what she called "technological normativity", which should not be confused, however, with so-called "legal normativity". Basically, technological normativity refers to the fact that every technology has a certain "normative" impact on our behavior insofar as it permits or facilitates us to do something (*regulative normativity*) and, at the same time, prohibits us from doing something else, thereby constraining our behavioral chances (*constitutive normativity*). We should thus acknowledge that modern law is embedded in certain practices which are shaped by pre-existing devices and tools. That is, modern law did not emerge in a technological vacuum, but it is a response to specific societal needs. However, as technological innovation brings new tools and devices into existence, we should acknowledge that modification of the legal framework is of urgent need to preserve those democratic values that our societies are – and continue to be – imbued with.

[13] A more exhaustive treatment of the relationship between ethics and technology is provided in Magnani [2007c]. Magnani discussed at length that morality is extended, and that recent technological advancements are clearly re-configuring some of the crucial aspects of our moral life. On this topic, Verbeek too points out that it is basically "a mistake to locate ethics exclusively in the 'social' realm of the human, and technology exclusively in the 'material' realm of the nonhuman" [Verbeek, 2009, p. 65]. Technologies are indeed part of our moral endowments in the way that our moral response can be shaped by technological devices.

precisely, through the activity of cognitive niche enrichment it delivers new kinds of affordances, which preexisting technologies could not furnish. This is the a creation of a kind of affordance that I call *adapting affordances*, as they exhibit *adaptability*.

Before I turn to issue related to affordance, for the rest of this chapter I shall discuss another important point: why and how cognitive niches may persist.

3.5   Cognitive Niche Maintenance and Group-Selection

3.5.1   *Cognitive Niche Maintenance*

I have just argued that cognitive niches contributes to introducing a second and non-genetic inheritance system insofar as the modifications brought about on the environment persist, and so be passed on from generation to generation. The main advantage of having this second inheritance system is that it enables humans to access a great variety of information and resources never personally experienced, but resulting from the activity of previous generations [Alvard, 2003]. That is, the information and knowledge humans can lean on are not simply transmitted, but they can be also *accumulated* in the human niches. Indeed, the knowledge I am talking about embraces a great variety of resources including knowledge about nature, social organization, technology, the human body, and so on.

Castro *et al.* [2004] have recently investigated the mechanism that favors the accumulation of knowledge beyond one generation. They argued a key factor enabling the growth of a cumulative inheritance system is the development of "the capacity to approve or disapprove their offspring's learned behavior". They argue that simply imitating what others are doing does not lead to accumulation of that particular trait being imitated. Imitation, and related forms of social learning, provide a general model for explaining how acquired traits can be transmitted from person to person, but not how they are accumulated into a kind of repertoire.[14]

Castro *et al.*'s intuition can be further developed in connection to the eco-cognitive framework I have described above. On one hand, a cognitive niche is a mediating structure enabling to store at various levels a great deal of information and resources; on the other hand, the capacity of approving or disapproving plays the major role of what I may call *cognitive niche maintenance*. All those activities of incorporation of a behavioral trait into the vast repertoire of resources delivered by a cognitive niche are examples of cognitive niche maintenance. The adoption or

---

[14] It is noteworthy that this contention seems to assign a fundamental role to morality for explaining the emergence of culture as a second inheritance system. Morality underlies all those activities of cognitive niche maintenance, because it favors certain ideas to persist at the expense of others. As a specific structure that connects rules and prescriptions with human emotional endowments [Adolphs, 2006], it is precisely related to the task of policing what ideas, behaviors, habits, etc. should be preserved, and what should not. In this sense, as suggested by several authors, it seems morality has characterized human communities for much of their evolutionary history [Boehm, 1999; Bingham, 2000; Wilson, 2002a; Wilson, 2002b; Magnani, 2009].

rejection of new learned traits should go under scrutiny and evaluation to preserve repertoire integrity.

The activities of cognitive niche maintenance can be carried out implicitly as well as explicitly. Either way cognitive niche maintenance is administered at group/coalition level. For instance, gossiping is certainly an activity of cognitive niche maintenance insofar as gossipers play the role of *assessors* policing group-serving behaviors and detecting cheaters [Wilson *et al.*, 2002]. In the section 1.3.2 I have already stressed how gossiping fallacies variously serve social control at the level of coalitions management. Generally speaking, the capacity of approve or disapprove acquired traits (behaviors, delegations to external objects, ideas, and so on) depends on the the presence of groups (or coalitions), which can only have the power for generating, sharing, and – most of all – *enforcing* public criteria and standards to approve and disapprove cultural traits.

Cognitive niche maintenance is a sort of conceptual bridge that permits us to connect the notion of niche construction to that of group-level selection: groups in their coalitionist nature can be regarded as adaptive units [Wilson, 2002b; Wilson, 2006]. I develop this contention arguing that a group can be smoothly considered an adaptive unit as far as its members occupy a cognitive niche, which locally alters the selective pressure, which they are subjected to. In this sense, although human communities rarely comprise only genetic relatives, group members can share what may be called phenotype familiarity or resemblance simply because they occupy the same ecological niche. In this sense, the notion of group-level adaptation can be fairly captured and properly addressed by the notion of niche construction as a key factor of evolution.

As it will be clear in the following, I am not indulging in any super-organism approach, which argues on the presence of group or collective mind. In our conception group-level functionalism acquires an explanatory power with respect to niche construction. That is, groups and coalitions function as integrated units insofar as they locally alter the selective pressure constructing or modifying the surroundings, namely, the cognitive niches. Individuals can exhibit a given set of cognitive behaviors, because they form part of a cognitive niche. In this sense, I agree with Wilson [2001], who clearly maintained group-level cognition approach does not necessary involve any ontological claim on the existence of super-organisms or group psychological traits – so to speak. My conception is akin to what he called the *social manifestation thesis*, according to which individuals can only have properties, which, however, are manifest "when those individuals form part of a group of a certain type" [Wilson, 2001, p. 265].

3.5.2 Finding Room for Group-Selection in Evolution

Here I will introduce some issues related to group selection.[15] I will try to understand what kind of role group selection may have in evolution, in the light of the niche construction approach I have developed in the previous section. My take is

---

[15] For a detailed treatment of this issue, see Okasha [2006].

that group selection plays a crucial role in evolution, insofar as it allows or facilitates a cognitive niche to persist, or not. This point is connected with the issue I treated in section 2.6 concerning the "maladaptive" dimension of biased rationality. In that section, I pointed out that the strategies based on competence-dependent information are successful as far as knowledge can persist and be accumulated and transmitted via the cognitive niche.

In literature, the issue concerning group selection arises in connection with those adaptations that are hard to explain referring only to *individual selection*. By definition an adaptation is a trait enhancing one's chance of survival and reproduction [Wilson and Wilson, 2007]. The paradox of traits like, for instance, altruism is that apparently they do not benefit those who demonstrate them. Quite the contrary, traits like altruism are commonly considered as *self-sacrificing*. If so, then they could hardly be thought of as the result of evolution, since they would sooner or later lead those bearing them to extinction.

This would not follow, if we assumed levels or units of selection beyond the individual unit, or the existence of a higher level of organization beyond the individual. This is basically the option set forth by David Sloan Wilson (cf. [Wilson, 1977; Wilson and Sober, 1994; Wilson, 2002b; Wilson, 2006; Wilson and Wilson, 2007]). Wilson's proposal is indeed attractive in as far as he put forward an interesting argument corroborated by mathematical tools, in which groups acquire an explanatory role for certain human behaviors. The case of altruism provides the clearest example to describe his approach to group selection.

Before getting onto Wilson's main line of argumentation, it is important to clarify what I mean by the term altruism. In literature we find two different, but apparently overlapping meanings [Stich, 2007]: *evolutionary altruism* and *psychological altruism*. The first case, refers to all those behaviors that are *group-serving* and insofar as a given behavior trades fitness from the individual level to the group, then it is altruistic. In the second case, the term altruism is attributed to those psychological dispositions or motivations underlying the decision to sacrifice a part of one's own benefit in favor of others. These two definitions of altruism are not in competition, but they do not necessarily involve one another. For instance, a certain behavior can be deemed as altruistic just because it contributes to the fitness transfer from individual to group level, no matter what the motivation, whether triggered by a selfish or an altruistic attitude. In the following I will refer to evolutionary altruism.[16]

Wilson's proposal starts from the very simple assumption that all organisms act so that their behavior has a certain impact on other organisms in a given population including on themselves. The problem of altruism arises when an organism opts for a certain behavior, which increases the fitness of one or more recipients while decreasing its own fitness. Consider, for instance, the case in which a given behavior x benefits everyone in a population by an amount $B(x)$ at a private cost of $C(x)$ so that we have:

$$B(x) > C(x)$$

---

[16] As becomes clear in the following, I will contend that the expression "altruistic behavior" could be replaced by that of "fitness-transfer behavior".

If one assumes that, *absolute individual fitness* is the appropriate criterion, which will predict who will be selected out then clearly, sooner or later, altruism will disappear because it is a disadvantageous trait for those who bear it. However, as Wilson pointed out [Wilson and Wilson, 2007], absolute individual fitness is the appropriate criterion, if we assume an *unstructured* population, in which groups are *randomly* formed. In fact, if the differences among groups are less than random – a condition technically called *positive assortment* – then absolute individual fitness looses its predictable power. An example can make this contention clearer.

Consider the case of a population, in which we have hunters and scroungers differently assorted in groups of the same size so that one group may have more hunters and the other more scroungers. Hunters benefit all the members of the group by an amount of 3 units at a private cost of 1; scroungers will benefit all the members of the group by an amount of 0 units at a private cost of 0. In this particular case, the overall fitness of each individual will not only be predicted by its absolute individual fitness, but also by the one *relative* to the group an individual is part of. In fact, it is easy to note that the probability of surviving and reproducing of each individual will increase, if it is be a member of the group in which there are more hunters. Under such a scenario, Wilson argued that *between*-group selection *overrides within*-group selection so that the appropriate criterion for predicting the fitness of each individual depends also on group fitness.

The approach of group or *multi-level* selection is supposed to explain the reason why altruism can spread and evolve.[17] In the example I illustrated, under specific conditions of positive assortment the fitness of the group depends on the presence of hunters, whose contribution to the group lessens the negative fitness they have at the absolute individual level.[18]

3.5.3   Group-Projecting Behaviors, Assortment, and the Stallation Hypothesis

The theoretical problem I are facing now is to discern what specific role group selection may have in evolution and whether or not it can be interpreted as a meaningful evolutionary force. The thesis I embrace is that group selection (or multilevel selection) is a fundamental ingredient for explaining certain adaptations, but it cannot be considered as a *unit* of selection.

Before proceeding it is worth distinguishing between two notions, which can lead to confusion. We distinguish *units of evolution* from *levels of selection*. As far as I am concerned here, a unit of evolution is any entity exhibiting *multiplication*, *variation*, and *heredity*, as Maynard Smith [1987] put it. The definition of "levels of selection" is a bit more complicated. As the example of relative individual fitness shows, a certain trait (i.e. altruism) can only be explained by assuming a higher

---

[17] A discussion about multi-level selection can be found in Field [2008].

[18] For a perspective contrasting this view, see Fletcher and Zwick [2004]. Fletcher and colleague explain why altruism may spread in randomly assorted groups.

level of selection. I posit that this is made possible by those behaviors that function as *fitness traders*.

The appearance and evolution of new level of selections is not something new in biology. Biological complexity results from major transition occurring during life history, involving levels of selections, which are structured in a hierarchy [Maynard-Smith and Szathmary, 1995; Michod, 2005]. From gene to gene networks, from unicellular to multicellular organism, life has followed a pattern of evolvability, in which groups became individuals bringing about new evolutionary units or a new level of selection [Michod and Aurora, 2003; Michod and Herron, 2006; Michod, 2007]. The transition from a lower to a higher level depends on fitness traders, which, in turn, result from cooperative interactions. The presence of cooperative interactions causes fitness transfer from one level to another.

Going back to human groups, I contend that they cannot be considered a new evolutionary unit or a new evolutionary individuality. Why? Because the fitness transferred from the individual level to the group level cannot be *inherited*. This is the major argument against the possibility of considering groups as units of evolution (or evolvability). As already mentioned, any collaborative behaviors cause a fitness transfer from the low level to the high level. The projecting nature of collaborative behavior is however limited within one generation, since there is no system of inheritance causing not only fitness transfer, but also *fitness inheritance*. A group cannot be interpreted as a stable unit, because it is constantly facing change [Ichinose and Arita, 2008]. After a number of generations, group composition completely changes. Besides, a group can face the process of dispersal, where a group as a unit can cease to exist for a number of reasons. Indeed, this is reasonable when assuming that a population is structured in different groups, in conflict with each other. However, groups are not divided by genetic relatedness, and therefore they can change over time, and even under go the aforementioned process of dispersal.

It is worth noting, however, that fitness transfer does not necessary imply that the transition toward the new unit of evolvability is completed. Therefore, I define behaviors such as altruism as *group-projecting behaviors*, meaning that groups are *projections* of an higher unit of evolution. That is, every time an individual behaves altruistically it acts as if the group actually exists. I maintain that such a definition would allow us to combine the intuitive force of Wilson's position without indulging in a *super-organicistic* approach to the matter.

The relevance of group-projecting behavior introduces quite a speculative issue, that is however useful in order to better understand the allegedly evolutionary meaning of groups. According to Stearns we are "stalled part way through a major evolutionary transition from individual to groups" [Stearns, 2007]. Although the hypothesis is indeed hard to prove, due to the very nature of the hypothesis itself, we have evidence, which makes it reasonable or – at least – not completely absurd and nonsensical. For instance, the astonishing degree of cooperation even among non-relatives, which characterizes human communities, the number of emotions and feelings underlying cooperative behaviors such as empathy and the sense of justice and duty, and the various social mechanisms to protect group members from cheating such as indirect reciprocity or second and third-party punishment.[19] The universal presence of morality in human communities is another clue, since morality is a powerful cultural artifact for composing within-group conflicts.

One of the interesting aspects of the stallation hypothesis is that social groups are the primary sources of trade-off mechanisms, which in theory are responsible for managing conflicts between two major behaviors, namely, self-serving behavior and group-serving behavior. Norms as policing factors are thought to regulate such mechanisms of conflict trade-off. However, it is worth noting that when positive assortment is strong, that is, when a group is stable, the adherence to social norms (including moral ones) becomes less urgent.[20] This is proof supporting the idea of the stallation hypothesis, since stability-dependent cooperation does not attribute a special adaptive role to cooperation *per se*, but to the tendency of assessing the optimal level of cooperation in a given situation. Conversely, morality requires a stronger commitment when group stability slows down or during the process of assorting.

There is another issue worth mentioning. According to the theory of human ecological dominance [Flinn *et al.*, 2005], the primary factor promoting or threatening an individual's reproductive success are *other people*. By the intense activity of niche construction, humans succeeded in altering their local environment so as to reduce threats coming from natural forces. This appears to attribute a selective role to groups with respect to other factors. In fact, as people owe much of their chance for survival and reproduction to the group they live in, competition between-groups becomes a primary source of selection.

3.5.4 An Eco-Cognitively Mediated Conception of Group Assortment

As already mentioned, one of the pre-requisites for completing a transition is fitness inheritance. That is, fitness variations emerging at group level should be inherited from past generations. That implies consideration of a group as a unit of evolution and evolvability. Indeed, human groups cannot be considered as units of evolution, because the fitness variations cannot be inherited. However, the presence of strong forms of group assortment is a major clue supporting the idea that we are part way through a major evolutionary transition. In human communities, group assortment manages group boundaries and membership over time. In this sense, all those assortative mechanisms, related to group boundaries and membership management, approximate what an inheritance system does when a major transition is completed. Here, two points should be stated clearly. First of all, cooperative behaviors cause a fitness transfer, which *projects* a new level of selection.[21] In turn, the group

---

[19] The so-called "social brain hypothesis" is discussed in Chapter 1.

[20] This idea is discussed in Lahti and Weinstein [2005] referring to the notion of *moral viscosity*. Moral viscosity is also treated in detail by Magnani [2011].

[21] As already noted above, the appearance of a new unit of evolution is not a pre-requisite to have a new level of selection.

projection mediates what is called the re-organization of fitness components. During a major transition re-organization of fitness components can be described as a process of *co-option* of pre-existing functions at individual level for designing new functions and traits serving at group level.

Now, the question is what kind of influence may a niche have on group assortment? Answering this question is crucial in order to integrate group selection into a more coherent framework for explaining the evolutionary impact of cognitive niches.

Pepper proposed a model in which *environmental feedback* is referred to as an alternate source of generating and maintaining group assortment [Pepper, 2000; Pepper, 2007]. In this case, such a model does not resort to the presence of any common descent or special cognitive abilities. Conversely, it simply assumes that a given trait alters the local environment so that organisms are prompted to react to the change following a certain pattern of response. For example, food supply may increase and thus benefit the group, as one responds altruistically to lower feeding efficiency. In this case, the altruistic trait causes feeding restraint, which in turn leads to a better distribution of food supply.

Starting from this idea, I argue that this approach may be fruitfully integrated within the theory of cognitive niche construction. In this case, I suggest assortment resorts to a set of *mediated* interactions that individuals living in the same cognitive niche share. As I will show in the next chapter, what they share is a set of affordances.

Let me resume the main points:

1. individual fitness relative to the group becomes the appropriate criterion to predict the evolutionary impact of a given trait under the condition of positive assortment;
2. group boundaries and membership rely on flexibility of the traits involved. Mechanisms of group assortment are the only means to guarantee stability in face of change and dispersal;
3. Mechanisms of group assortment permit a group to have an approximating system of inheritance;
4. A cognitive niche has an influence on group assortment in terms of environmental feedback.

In conclusion, relatedness is basically the degree to which benefits due to a given behavior are restricted to the members of one group. Adopting an eco-cognitively mediated conception of assortment simply means that the benefits are restricted to those who occupy the same cognitive niche.

3.6   Concluding Remarks

In this chapter I have illustrated how the notion of the cognitive niche plays a crucial role for developing the idea of moving the bounds to rationality. This is an attempt to view cognition from a broader perspective able to give an evolutionary dimension to it. As a matter of fact, humans and many other organisms continuously manipulate the environment in order to take advantage of it. In doing this, they are engaged in a process of altering or even creating external structures to lessen and overcome their limits. New ways of coping with the environment, through both evolution and cultural evolution (i.e "cognitive niche construction") are thus created.

In the last part of the chapter, I dealt with the issue related to cognitive niche maintenance. I have clarified some controversial points concerning group selection. The idea I introduced is that a cognitive niche may persist, and thus acquire evolutionary meaning, insofar as there is a group maintaining it. In turn, cognitive niche maintenance is favored by various activities of group assortment, which I have interpreted as eco-cognitively mediated ones.

The next chapter will put forward an alternative conception of the cognitive niche, based on the notion of affordance. As I will demonstrate, a cognitive niche can be defined as a set of affordances, which are basically environmental chances that can be found or manufactured upon occasion. The introduction of this definition is thought to solve some of the problems concerning what should be counted as a cognitive niche.

Chapter 4
Building Cognitive Niches: The Role of Affordances

Introduction

In the second part of Chapter 2, I pointed out that rationality is *un-biased*, as humans manipulate their environment to build up various external structures which become increasingly *symptomatic* in relation to what is happening in their environment. The recourse to external structures that are more symptomatic makes biased rationality less and less appealing, since the new option appears to be better grounded in experience. This chapter will aim at illustrating how humans manipulate their local environment in order to *move the bounds* of rationality.

The notion of affordance will provide the suitable conceptual framework to illustrate the construction of ever more symptomatic external structures for making decisions and solving problems. By manipulating the environment, namely, by means of niche construction, humans unearth new chances that, in turn, contribute to moving the bounds of rationality. Chances are not simply information, but they are "affordances", namely, *environmental anchors* that allow us to better exploit external resources.

In Chapter 3 I illustrated the theory of niche construction applied to cognition. I introduced the notion of cognitive niche construction referring to the fact that the designing activities which humans lean on to manipulate the environment are part of human cognition insofar as they unearth additional chances for behavior control. Thus, human cognition and its evolutionary dimension can be better understood in terms of environmental situatedness where information and resources are not only given, but they are actively sought and even manufactured. This is basically the idea of human cognition as a chance-seeking system.

After presenting some of the conceptual muddles affordance suffers from, I introduce the notion of abduction in order to clear up any ambiguities and misconceptions still present in current debate. Going beyond a mere *sentential* conception, I will argue that the role played by abduction is two fold. First of all, it decisively leads us to a better definition of affordance. As I will discuss in detail in section 4.4, I will posit that affordances can be related to the variable (degree of) *abductivity* of a configuration of signs. Secondly, abduction turns out to be a valuable candidate for solving the problem related to whether affordance detection is mediated or not.

E. Bardone: Seeking Chances, COSMOS 13, pp. 77–100, 2011.
springerlink.com © Springer-Verlag Berlin Heidelberg 2011

In closing section 4.5, I will come back to the notion of cognitive niche enrichment introduced in Chapter 3. First of all, I will introduce the notion of adapting affordance. Secondly, I will show how conceptually rich that notion is in assessing the eco-cognitive discontinuity that Ambient Intelligence is promising to bring about.

4.1   Cognitive Niche as a Set of Affordances

4.1.1   The Notion of Affordance

One of the most disturbing problems with the notion of affordance is that any examples provide different, and sometimes ambiguous insights on it. This fact makes very hard to give a conceptual account of it. That is to say, when making examples everybody grasps the meaning, but as soon as one tries to conceptualize it the clear idea one got from it immediately disappears. Therefore, I hope to go back to examples from abstraction without loosing the intuitive simplicity that such examples provide to the intuitive notion.

The entire debate during the last fifteen years about the notion of affordance is very rich and complicated, but also full of conflicts and ambiguities. This subsection aims at giving just an overview of some issues I consider central to introduce to my treatment.

Gibson defines "affordance" as what the environment offers, provides, or furnishes. For instance, a chair affords an opportunity for sitting, air breathing, water swimming, stairs climbing, and so on. By cutting across the subjective/objective frontier, affordances refer to the idea of agent-environment mutuality. Gibson did not only provide clear examples, but also a list of definitions [Wells, 2002] that may contribute to generating possible misunderstanding:

1. affordances are opportunities for action;
2. affordances are the values and meanings of things which can be directly perceived;
3. affordances are ecological facts;
4. affordances are distributed representations.

I contend that the Gibsonian ecological perspective originally achieves two important results. First of all, human and animal agencies are somehow hybrid, in the sense that they strongly rely on the environment and on what it offers. Secondly, Gibson provides a general framework about how organisms directly perceive objects and their affordances. His hypothesis is highly stimulating: "[...] the perceiving of an affordance is not a process of perceiving a value-free physical object [...] it is a process of perceiving a value-rich ecological object", and then, "physics may be value free, but ecology is not" [Gibson, 1979, p. 140]. These two issues are related, although some authors seem to have disregarded their complementary nature. It is important here to clearly show how these two issues can be considered two faces of the same medal. Let us start our discussion.

4.1.2 Affordances as Action Opportunities

Several authors have been extensively puzzled by the claim repeatedly made by Gibson that "an affordance of an object is directly perceived" [Greeno, 1994; Stoffregen, 2003; Scarantino, 2003; Chemero, 2003]. During the last few years an increasing number of contributions has extensively debated the nature of affordance as opportunity for action. Consider for instance the example "stairs afford climbing". In this example, stairs provide us with the opportunity of climbing; we climb stairs because we perceive the property of "climbability", and that affordance emerges in the interaction between the perceiver and stairs [Chemero, 2003; Stoffregen, 2003]. In order to prevent from any possible misunderstanding, it is worth distinguishing between "affordance property" and "what" and object affords [Natsoulas, 2004]. In the former sense, the system "stairs-plus-perceiver" exhibits the property of climbability, which is an *affordance property*. Whereas in the latter the possibility of climbing is clearly *what* an object affords.

4.1.3 Affordances as Ecological Facts

Concerning this point, Gibson argued that affordances are ecological facts. Consider, for instance, a block of ice. Indeed, from the perspective of physics a block of ice melting does not cease to exist. It simply changes its state from solid to liquid. Conversely, to humans a block of ice melting does go out of existence, since that drastically changes the way we can interact with it. A block of ice can chill a drink the way cold water cannot. Now, the point made by Gibson is that we may provide alternative descriptions of the world: the one specified by affordances represents the environment in terms of action possibilities. As Vicente put it, affordances "[...] are a way of measuring or representing the environment with respect to the action capabilities of an individual [...] one can also describe it [a chair] with respect to the possibilities for action that it offers to an organism with certain capabilities" [Vicente, 2003]. Taking a step further, I may claim that affordances are chances that are *ecologically rooted*. They are ecological rooted because they rely on the mutuality between an agent (or a perceiver) and the environment. As ecological chances, affordances are the result of a hybridizing process in which the perceiver meets the environment. The emphasized stress on the mutuality between the perceiver and the environment provides a clear evidence of this point.

4.1.4 Affordances as Distributed Representations

Recently, Zhang and Patel [2006], also going beyond the ecological concept of affordance in animals and wild settings by involving its role in human cognition and artifacts, in an unorthodox perspective, connect the notion of affordance to that of distributed representation. They maintain that affordances can be also related to the role of distributed representations extended across the environment and the organism. These kinds of representation come about as the result of a blending process between two different domains: on one hand the internal representation space, that is the physical structure of an organism (biological, perceptual, and cognitive faculties); on the other the external representation of space, namely, the structure of the environment and the information it provides. Both these two domains are described by constraints so that the blend consists of the allowable actions. Consider the example of an artifact like a chair. On one hand the human body constrains the actions one can make; on the other the chair has its constraints as well, for instance, its shape, weight, and so on. The blend consists of the allowable actions given both *internal* and *external* constraints.[1]

Patel and Zhang's idea tries to clarify that affordances result from a hybridizing process in which the environmental features and the agent's ones in terms of constraints are blended into a new domain which they call *affordance space*. Taking a step further, Patel and Zhang define affordances as allowable actions. If this approach certainly acknowledges the hybrid character of affordance I have described above and the mutuality between the perceiver and the environment, it seems however lacking with regard to its conceptual counterpart. As already argued, affordances are action-based opportunities.

4.1.5 Affordances as Evolving Interactional Structures

Taking advantage of some recent results in the areas of distributed and animal cognition, I can find that a very important aspect that is not sufficiently stressed in literature is the dynamic one, related to designing affordances, with respect to their evolutionary framework: human and non-human animals can "modify" or "create" affordances by manipulating their cognitive niches. Moreover, it is obvious to note that human, biological bodies themselves evolve: and so I can guess that even the more basic and wired perceptive affordances available to our ancestors were very different from the present ones.[2] Of course different affordances can also be

---

[1] This idea can also be connected to the concept of cognitive fluidity argued by Mithen [1996]. From the perspective of cognitive palaeoanthropology, Mithen claimed that the modern mind is characterized by the capacity of applying to heterogeneous domains forms of thinking originally designed for specific tasks. He also contends that in hominids this change originated through a blend of two different intelligence domains, namely, that of internal representations entities and external artifacts [Magnani, 2006a].

[2] The term "wired" can be easily misunderstood. Generally speaking, I accept the distinction between cognitive aspects that are "hardwired" and those which are simply "pre-wired". By the former term I refer to those aspects of cognition which are fixed in advance and not modifiable. Conversely, the latter term refers to those abilities that are built-in prior the experience, but that are modifiable in later individual development and through the process of attunement to relevant environmental cues: the importance of development, and its relation with plasticity, is clearly captured thanks to the above distinction. Not all aspects of cognition are pre-determined by genes and hard-wired components.

detected in children, and in the whole realm of animals. I will come back to this issue in section 4.4.

4.2 The Two Views on Affordance: The Ecological and the Constructivist Approach at Stake

4.2.1 The Two Views

The theory of affordance potentially re-conceptualizes the traditional view of the relationship between action and perception according to which we extract from the environment those information which build up the mental representation that in turn guides action [Marr, 1982]. From an ecological perspective, the distinction between action and perception is questioned. The notion of affordance contributes to shed light on that issue fairly expanding it.

I posit that the Gibsonian ecological perspective originally achieves two important results. First of all, human and animal agencies are somehow hybrid, in the sense that they strongly rely on the environment and on what it offers. Secondly, Gibson provides a general framework about how organisms directly perceive objects and their affordances. His hypothesis is highly stimulating: "[...] the perceiving of an affordance is not a process of perceiving a value-free physical object [...] it is a process of perceiving a value-rich ecological object", and then, "physics may be value free, but ecology is not" [Gibson, 1979, p. 140]. These two issues are related, although some authors seem to have disregarded their complementary nature. It is important here to clearly show how these two issues can be considered two faces of the same medal.

We may provide alternative descriptions of the world: the one specified by affordances represents the environment in terms of action possibilities. I have already cited Vicente [2003] arguing that affordances a way of measuring or representing the environment with respect to the action capabilities of an individual. Taking a step further, I may claim that affordances are chances that are *ecologically rooted*. They are ecological rooted because they rely on the mutuality between an agent (or a perceiver) and the environment. As ecological chances, affordances are the result of a hybridizing process in which the perceiver meets the environment. The emphasized stress on the mutuality between the perceiver and the environment provides a clear evidence of this point.

In his research Gibson basically referred to "direct" perception, which does not require the internal inferential mediation or processing by the agent. Donald Norman from Human Computer Interaction studies challenged the original Gibsonian notion of affordance also involving mental/internal processing: "I believe that affordances result from the mental interpretation of things, based on our past knowledge and experience applied to our perception of the things about us" [Norman, 1988, p. 14]. It appears clear that in this case affordances depend on the organism's experience, learning, and full cognitive abilities, i. e. they are not independent of them, like Gibson maintained. For example infants 12 to 22 weeks old already show complicated cognitive abilities of this type, as reported by Rader and Vaughn [2000]. These abilities allow them to lean on prior experience of an object and therefore detect what Rader and Vaughn call "hidden affordances". As argued by these authors, hidden affordances are those affordances specified by the information not available at the time of the interaction, but drawn from past experiences [Rader and Vaughn, 2000, p. 539]. The same event or place can have different affordances to different organisms but also multiple affordances to the same organism. Following D. Norman's perspective, affordances suggest a range of *chances*: given the fact that artifacts are complex things and their affordances normally require high-level supporting information, it is more fruitful to study them following this view.

To give an example, perceiving the full range of the affordances of a door requires complex information about for example direction of opening or about its particular pull. Becoming attuned to invariants and disturbances often goes beyond the mere Gibsonian direct perception and higher representational and mental processes of thinking/learning have to be involved.[3] This means that for example in designing an artifact to the aim of properly and usefully exhibiting its full range of affordances I have to clearly distinguish among two levels: 1) the construction of the utility of the object and 2) the delineation of the possible (and correct) perceptual information/cues that define the available affordances of the artifact. They can be more or less easily be undertaken by the user/agent [Gaver, 1991; Warren, 1995; McGrenere and Ho, 2000]: "In general, when the apparent affordances of an artifact match its intended use, the artifact is easy to operate. When apparent affordances suggest different actions than those for which the object is designed, errors are common and signs are necessary" [Gaver, 1991, p. 80]. In this last case affordances are apparent because they are simply "not seen". In this sense information arbitrate the perceivability of affordances, and we know that available information often goes beyond what it can be provided by direct perception but instead involves higher cognitive endowments.

Vicente contends that it has to be said that of course it is impossible to think that direct perception can explain all psychological phenomena, like mainly Gibsonian researchers seem to maintain. Moreover, according to Reed, the opinion that mediated perception or cognition is inconsistent with Gibson's view of ecological psychology is "simply mistaken" [Reed, 1988, p. 305], like the following passage by Gibson would clearly illustrate:

> At least three separate levels [of theorizing] will be required: first, a theory of how we perceive the surfaces of objects [...]; second, a theory of how we perceive representations, pictures, displays, and diagrams; and third, a theory of how we apprehend symbols. There is no reason to suppose that the physiological concomitants of all these experiences will be the same; in fact, since pictures and symbols presuppose objects, their physiological explanations will probably have to be found at increasing levels of complexity [Gibson, 1951, p. 413].

---

[3] Some authors [Turvey and Shaw, 2001; Hammond *et al.*, 1987] pointed out that high-level organisms' cognitive processes like those referred to language, inference, learning, and the use of symbols would have to be accounted for by a mature ecological psychology.

Of course Gibson manly preferred to study the first of the three categories of theories, related to relatively narrow psychological phenomena.

4.2.2 Confronting the Evidences

The hypothesis that the representation we have of the world includes also ecological chances for action, namely affordances, has been investigated from a neuro-anatomical perspective. Although neuro-anatomical evidences do not provide a conclusive argument, they have contributed to shed light on some aspects about affordance that otherwise would have remained highly hypothetical or, at least, based on partial conjectures.

It is J. Norman who, taking advantage of a wholly neuropsychological perspective, tries to account for a reconciliation of the two approaches above (ecological and constructivist). They resort to two cortical visual systems, the first of which he calls dorsal – hardwired, direct and active, less representational, without the recourse to memory, and so expressing Gibson's affordances – and the second ventral, which is more representational and judgmental, indirect, and related to mentalistic processes, and which basically performs different transformations of the available visual information. Both systems perform different functions, and, present consciousness at different degrees (for example the ventral system brings the relevant information, picked up in a more unconscious way by the dorsal system, to conscious awareness). Finally, both systems analyze the visual input, but the analysis is carried out for different purposes. Certainly both systems deal with object shapes, sizes, and distances, but

> The primary function of the ventral system would seem the *recognition* and *identification* of the visual input. Recognition and identification must depend on some comparison with some stored representations. In contrast, the primary function of the dorsal system is analysis of the visual input in order to allow *visually guided behavior* vis à vis the environment and objects in it (e.g. painting, reaching, grasping, walking towards or through, climbing, etc.). [...] Thus, when one picks up a hammer, the control and monitoring of the actual movements is by the dorsal system but there also occurs intervention of the ventral system that recognizes the hammer as such and directs the movement towards picking up the hammer by the handle and not by the head [Norman, 2002, p. 84].

The two visual systems are highly integrated, and one may say that the distinction works at an anatomical level. But, if highly integrated, how does the transition between one another come about? How to explain the switch between the motor processing and the visual processing? Young [2006] questioned Norman's view according to which there would be an alignment of affordances with the dorsal stream. He claimed that only certain affordances are processed along the dorsal stream. This seems to suggest a more complicated classification of affordances depending upon their respective neurological underpinnings, but also – I would add – their cognitive meaning. Recent studies on patients affected by brain damage that impaired dorsal or ventral stream have contributed to shed light on that issue.[4]

Consider for instance *visual agnosia*. Visual agnosia is caused by damage of the ventral stream of one's visual system [Milner and Goodale, 1995], which impedes patients to consciously experience objects and access semantic knowledge related to them. Empirical studies reported that patients suffering from visual agnosia are still able to perform certain tasks that require the detection of simple affordances, for instance, holding and grasping pliers. However, the same patients showed the inability of understanding how to use instruments to accomplish more skilled tasks, for example, clasping the handles to manipulate jaws [Carey *et al.*, 1996]. As argued by Young [2006], that inability may be due to the lack of *functional knowledge* required to skillfully manipulate objects, which heavily depends on previous experiences. The role of information provided by past experience suggests the involvement of ventral stream in detecting more complex affordances as reported by Milner [2001]. Consider now the case of *optic ataxia* (dorsal stream impaired). Patients suffering from optic ataxia were able to pantomime a grasp of an object, if seen earlier, and thus retained in memory. They overcame their visuomotor deficit by relying on an off-line guidance neurologically based on the ventral stream, which, in turn, provides memory-based information.

What they lack is therefore the ability to unconsciously adjust ongoing movements that seems to suggest the existence of an automatic pilot, which in this case results impaired [Himmelbach *et al.*, 2006, p. 2750]. That is, patients successfully interact with the object by retrieving past information stored in their memory [Milner *et al.*, 2001] instead of picking them up upon occasion. As put it by [Himmelbach and Karnath, 2005, p. 633], "[...] the contribution of ventral system increases as the delay between target presentation and movement execution gets longer". This introduces a division of labor between dorsal and ventral system in allowing people to interact with their environment. From this perspective affordances processed along the dorsal stream are merely picked up unconsciously, "whereas those processed via the ventral stream constitute one aspect of the content of the subject's phenomenal experience, although I accept that this does not mean that the subject must be reflectively aware of such content" [Young, 2006, p. 141].

Taking a step further, I may argue that the ventral stream acquires importance when detecting affordances relies more on functional knowledge just like the case of hidden affordances I previously introduced. Here, detecting hidden affordances is the result of an inferential/mental activity which clearly relies on semantic

---

[4] From a methodological perspective, the so-called "double- dissociation of function" constitutes a major evidence for the existence of neurologically distinct functional systems. Generally speaking, double dissociation is considered a strong neurological evidence when a lesion of structure X [ventral stream] will specifically disrupt function A [functional knowledge] while sparing function B [manipulation of an object], and a lesion of structure Y [dorsal stream] will specifically affect function B [manipulation of an object] while function A would remain intact [functional knowledge]. For a critical point of view on double-dissociation and the two visual systems, see Pisella *et al.* [Pisella *et al.*, 2006].

information (functional knowledge) that come from past experience and, indeed, depend on a learning process.[5]

Indeed, people may even make automatic the process of detecting a hidden affordance, once they are got used to it. For instance, patients with lesions of the ventral stream have preserved ability to make gestures in response to visually presented objects. However, this happens only with familiar objects [Decety and Grèzes, 2006]. Conversely, when presented unfamiliar objects patients with the ventral stream impaired are unable to perform a correct grasp. In this case, affordances are not readily available, but they require an additional mediation that in those patients cannot be activated, since the ventral stream results impaired. This seems to suggest that familiarity, and thus ontogenetic process like learning, can really make the difference in understanding the division of labor between ventral and dorsal stream. Moreover, it contributes to shed light on the evolutionary dimension of affordance that has been often discarded. I will come back to this point in section 4.3.

4.3   The Breadth of Abductive Cognition

A hundred years ago, Peirce [1931–1958] coined the concept of abduction in order to illustrate that the process of scientific discovery is not irrational and that a methodology of discovery is possible. Peirce interpreted abduction essentially as an "inferential" *creative process* of generating a new hypothesis. Abduction has a logical form (fallacious, if we model abduction by using classical syllogistic logic) distinct from deduction and induction. Reasoning which starts from reasons and looks for consequences is called *deduction*; that which starts from consequences and looks for reasons is called *abduction*.

Magnani [2001] defines abduction as the process of *inferring* certain facts and/or laws and hypotheses that render some sentences plausible, that *explain* or *discover* some (eventually new) phenomenon or observation; it is the process of reasoning in which explanatory hypotheses are formed and evaluated. An example of abduction is the method of inquiring employed by detectives: in this case we do not have direct experience of what we are taking about. Say, we did not see the murderer killing the victim. But we infer that given certain signs or clues, a given fact must have happened. More generally, we guess a hypothesis that imposes order on data.

According to Magnani [2001], there are two main epistemological meanings of the word abduction: 1) abduction that only generates "plausible" hypotheses ("selective" or "creative") and 2) abduction considered as inference "to the best explanation", which also evaluates hypotheses. An illustration from the field of medical knowledge is represented by the discovery of a new disease and the manifestations

---

[5] Also Gibson suggested, as I have already pointed out, that "the process of pickup is postulated to be very susceptible to development and learning. The opportunities for educating attention, for exploring and adjusting, for extracting are unlimited. The increasing capacity of a perceptual system to pick up information, however, does not in itself constitute information"[Gibson, 1979, p. 250].

it causes which can be considered as the result of a creative abductive inference. Therefore, "creative" abduction deals with the whole field of the growth of scientific knowledge. This is irrelevant in medical *diagnosis* where instead the task is to "select" from an encyclopedia of pre-stored diagnostic entities. We can call both inferences ampliative, selective and creative, because in both cases the reasoning involved amplifies, or goes beyond, the information incorporated in the premises [Magnani, 1992b].

Abduction can fairly account for some crucial theoretical aspects of hypothesis generation as well as manipulative ones. Accordingly, Magnani [2001] distinguishes between two general abductions, *theoretical abduction* and *manipulative abduction*. *Theoretical abduction* illustrates much of what is important in creative abductive reasoning, in humans and in computational programs. It regards verbal/symbolic inferences, but also all those inferential processes which are model-based and related to the exploitation of internalized models of diagrams, pictures, etc.

Theoretical abduction does not account for all those processes of hypothesis generation relying on a kind of "discovering through doing", in which new and still unexpressed information is codified by means of manipulations of some external objects (*epistemic mediators*). These inferential processes are defined by manipulative abduction which, conversely, is the process of inferring new hypotheses or explanation occurring when the exploitation of environment is crucial. More generally, manipulative abduction occurs when many external things, usually inert from the semiotic point of view, can be transformed into "cognitive mediators" that give rise - for instance in the case of scientific reasoning - to new signs, new chances for interpretants, and new interpretations.

Therefore, manipulative abduction represents a kind of redistribution of the epistemic and cognitive effort to manage objects and information that cannot be immediately represented or found internally (for example exploiting the resources of visual imagery).[6] If the structures of the environment play such an important role in shaping our semiotic representations and, hence, our cognitive processes, we can expect that physical manipulations of the environment receive a cognitive relevance.

This distinction contributes to going beyond a conception of abduction, which is merely logical, say, related only to its sentential and computational dimension, towards a broader semiotic dimension worth investigating. Peirce himself fairly noted that the all thinking is in signs, and signs can be icons, indices, or symbols. In this sense, all *inference* is a form of sign activity, where the word sign includes "feeling, image, conception, and other representation" [Peirce, 1931–1958, 5.283]. This last consideration clearly depicts the semiotic dimension of abduction, which will be crucial in deepening the relationship between abduction and affordance.

I strongly agree with Magnani [2009] who posits that abduction contributes to shedding light on a wide range of phenomena – from perception to higher forms

---

[6] It is difficult to preserve precise spatial relationships using mental imagery, especially when one set of them has to be moved relative to another.

of cognition – which otherwise would not be appropriately considered and under-stood. In the previous section I illustrated the problem with affordances, whether they are directly perceived or mediated by higher cognitive processes. Abduction clarifies this issue in the sense that it goes beyond the traditional conception, which contrasts automatic or spontaneous response and perception, on one hand, with me-diated inferences and more plastic and reasoned forms of cognition, on the other. Apparently, the difference does not seem to reside in the underlying processes (me-diated or not), but the different cognitive endowments organisms (humans included) can take advantage and make use of. In this sense, it is better to distinguish two different aspects of the same problem. First of all, the word inference is not ex-hausted by its logical/higher cognitive aspects but also refers to the effect of various sensorial activities [Magnani, 2009]. Secondly the ability to make use of various cognitive endowments (instinctual or plastic), which is ultimately connected to ab-duction, is an evolving property. That is, it is open to improvements, modification, and evolution.

Perceptual judgment is the best example to illustrate that it is not only conscious abstract thought that we can consider inferential (cf. Magnani [2009]). Indeed, per-ceptual judgment is meant to be a spontaneous and automatic response, which is connected to various cognitive endowments, which are "instinctual" and hard-wired by evolution. Seeing is meant to be a direct form of cognition, which does not need any kind of mediation, since it happens without thinking – almost instantaneously. However, what about the case in which stimuli appear to be ambiguous, say, they need to be disambiguated (for instance, a face which is half familiar half not)? Usu-ally, as we get more and more clues, we get a clearer picture about who is the person approaching us. That is symptomatic of the fact that perceptual judgment is a sign-activity, namely, abduction. Indeed the perceptions we have cannot be deliberately controlled the way scientific inferences are. We do not have any conscious access to them. However, perceptions are always withdrawable, just like the case we are presenting, our perceptual judgment can be subjected to changes and modifications, as we acquire more clues/signs. In this sense, I may say that what we *see* is what our visual apparatus can, "explain" [Rock, 1982; Thagard, 1988; Hoffman, 1998; Magnani, 2001].[7] That is, people are very got used to impose order on various, even ambiguous, stimuli, which can be considered "premises" of the involved ab-duction [Magnani, 2001, p. 107]. More generally, those forms of cognition just like perceptual judgment are still abduction, even though, as brilliantly Peirce noted, they tend "to obliterate all recognition of the uninteresting and complex premises" from which they was derived [Peirce, 1931–1958, 7.37]. In this sense, percep-tual judgment is only apparently a not-mediated process: it is semi-encapsulated in the sense that it is not insulated from "knowledge" (cf. Raftopoulos [2001a; 2001b] and Magnani [2001; 2009]).

---

[7] Recently, Thagard [2010] claimed that perception cannot be described as abductive. He argued that perception involves some neuropsychological processes that do not fit with the definition of abduction as the process of generation and evaluation of explanatory hypotheses.

Another example supporting the semi-encapsulated nature of perception is provided by those perceptual judgments made by the experts.[8] Consider a professional meteorologist. What a trained meteorologist sees when looking at the sky clearly goes beyond what ordinary people see. They just see clouds, whereas a meteorologist would see various types and subtypes of clouds (cirrus clouds, altocumulus clouds, and many more) revealing a great deal about how the weather will evolve over the course of the next few hours. Thanks to the training they received, meteorologists as experts are able to spontaneously impose order over an additional range of signs coming from the sky, which are *learnt* and *knowledge-dependent*. Once familiar with a particular set of signs, then perceptual judgment proceeds automatically and with no further testing, although this may require a long period of study and training if sophisticated knowledge is involved.

4.4   Affordances as Abductive Anchors: Going beyond the Two Views

As already mentioned, Gibson defines affordance as what the environment offers, provides, or furnishes. For instance, a chair affords an opportunity for sitting, air breathing, water swimming, stairs climbing, and so on. But what does that exactly mean from an abductive perspective I introduced so far?

Within an abductive framework, that a chair affords sitting means we can perceive some clues (robustness, rigidity, flatness) from which a person can easily say "I can sit down". Now, suppose the same person has another object O, and she/he can only perceive its flatness. He/she does not know if it is rigid and robust, for instance. Anyway, he/she decides to sit down on it and he/she does that successfully. Is there any difference between the two cases?

I claim the two cases should be distinguished: in the first one, the cues we come up with (flatness, robustness, rigidity) are *highly diagnostic* to know whether or not we can sit down on it, whereas in the second case we eventually decide to sit down, but we do not have any precise clue about. How many things are there that are flat, but one cannot sit down on? A nail head is flat, but it is not useful for sitting. That is to say, that in the case of the chair, the signs we come up with are "highly diagnostic". That is, affordances can be related to the variable (degree of) *abductivity* of a configuration of signs: *a chair affords sitting* in the sense that the action of sitting is a result of a sign activity in which we perceive some physical properties (flatness, rigidity, etc.), and therefore we can ordinarily "infer" (in Peircean sense) that a possible way to cope with a chair is sitting on it.

More generally, the original Gibsonian notion of affordance deals with those situations in which the signs and clues we can detect prompt or suggest a

---

[8] The so-called visual abductions [Magnani, 2001] are essential also in science in which new and interesting discoveries are generated through some kind of visual-model based reasoning. On this topic, see for instance Shelley [1996] and Gooding [2004]. Bertolotti and Magnani [2010] discuss visual abduction in the generation of beliefs in super-natural agents.

certain action (or exploitation) rather than others. In this sense, I maintain that finding/constructing affordances deals with a (semiotic) inferential activity [Windsor, 2004; Magnani, 2009]. Indeed, we may be afforded by the environment, if we can detect those signs and cues from which we may abduce the presence of a given affordance [Magnani and Bardone, 2008].

There are a number of points I should now make clear. They will help us clarify the idea of affordances as abductive anchors. First of all, an affordance can be considered as a *hypothetical* sign configuration. An affordance is neither the result of an abduction nor the clues (in Gibsonian terms, the information specifying the affordance). I contend that an affordance informs us about an *environmental symptomaticity*, meaning that through a hypothetical process we recognize that the environment suggests for us and, at the same time, enables us to behave a certain way. We can do something with the environment, we can have a certain interaction, we can exploit the resources ecologically available in a certain way. This is particularly interesting if we go back to the issue I introduced in the first section about how to transform the environment from a source of constraints to a source of resources. In that section I pointed out that the human agent (like any other living organism) tries to attain a stable and functional relationship with their surroundings. I now claim that affordances *invite* us to couple with the environment by informing us about possible symptomaticities. In doing so, affordances become *anchors* transforming the environment into an *abductive texture* that helps us establish and maintain a functional relationship with it.

The second point worth mentioning is related to how the human agent regulates his relationship with his environment. I have partly answered this question. My idea is that the human agent *abductively* regulates his relationship with the environment. That is, the human agent is constantly engaged in controlling his own behavior through continuous manipulative activity. Such manipulative activity (which is eco-cognitive one) hangs on to abductive anchors, namely, affordances that permit the human agent to take some part of the environment as local representatives of some other. So, the human agent operates in the presence of abductive anchors, namely, affordances, that stabilize environmental uncertainties by directly signaling some pre-associations between the human agent and the environment (or part of it).

The third point is that an affordance should not be confused with a resource. I have just argued that an affordance is what informs us that the environment may support a certain action so that a resource can be exploited. Going back to the example of the chair, I contend that an affordance is what informs us that we can perform a certain action in the environment (*sitting*) in order to exploit part of it as a resource (*the chair*). This is coherent with the idea introduced by Gibson and later developed by some other authors who see an affordance as an "action possibility" (see section 4.1.2).

The idea that an affordance is not a resource but rather, something that offers information about one, allows it to be seen as anything involving some eco-cognitive dimension. That is, insofar as we gain information on *environmental symptomaticity* environmental symptomaticity to exploit a latent resource, then we have an affordance.

Generally, speaking, being or not being afforded by external objects is something related to:

1. the various instinctual or hard-wired endowments *hardwired behavior* humans are already attuned to,
   and
2. the various plastic cognitive endowments *prewired behavior* they have attuned to by learning or, more generally, they can make use of by using past knowledge and/or more sophisticated internal*plasticity* operations.

In the first case, affordances are already available and belong to the normality of the adaptation of an organism to a given ecological niche. Thus, in most cases detecting an affordance is a spontaneous abduction because this chance is already present in the perceptual endowments of human and non-human animals. That is to say, there are affordances that are somehow pre-wired and thus neurally pre-specified like action codes which are activated automatically by visual stimuli. This is, indeed, coherent to what Gibsonian conception refers to.

Organisms have at their disposal a standard (instinctual) endowment of affordances (for instance through their wired sensory system), but at the same time they can extend and modify the range of what can afford them through the appropriate cognitive abductive skills (which are more or less sophisticated). That is to say, some affordance relies on prior knowledge, which cannot be available at the time of the interaction, as already mentioned.

The fact that complex affordances would require the appropriate cognitive skills and knowledge to be detected does not mean to say they cease to be affordances, strictly speaking. As pointed out in section 4.2.2, familiarity is a key component to the involvement of the dorsal stream, which makes it easier to detect affordances. The term "familiarity" covers a wide range of situations in which affordance detection becomes almost automatic – coherently with the Gibsonian view – even if it requires an additional mediation of previous knowledge and even training.

That is the case of experts. Experts take advantage of their advanced knowledge within a specific domain to detect signs and clues that ordinary people cannot detect. Here again, a patient affected by pneumonia affords a physician in a completely different way compared with that of any other non medical person. Being abductive, the process of perceiving affordances mostly relies on a continuous activity of hypothesizing which is cognition-related. That A affords B to C can also be considered from a semiotic perspective as follows: A signifies B to C. A is a sign, B the object signified, and C the interpretant. Having cognitive skills (for example knowledge content and inferential capacities but also suitable wired sensory endowments) in a certain domain enables the interpretant to perform certain abductive inferences from signs (namely, perceiving affordances) that are not available to people not possessing those tools. To ordinary people a cough or chest pain are not diagnostic, because they do not know what the symptoms of pneumonia or other diseases related to cough and chest pain are. Thus, they cannot make any abductive inference of this kind.

Humans may take advantage of additional affordances, which are not *built-in*, so to speak, but have become stabilized. Now one important point should be added in. The possibility of displaying certain abductive skills depends also on those plastic endowments, which have been created and manufactured for a specific purpose. In this sense, the possibility of creating and then stabilizing affordances relies on the various transmission and inheritance systems that humans (but also other non-human creatures) display. That is, affordances are stabilized with recourse to a variety of means using, for instance, genetic but also behavioral and/or symbolic helpers. Environments change and so do perceptive capacities, when enriched through new or higher-level cognitive skills, which go beyond the ones granted by merely instinctual levels.

This dynamics explains the fact that if affordances are usually stabilized this does not mean they cannot be modified and changed and that new ones can be formed. Affordances are also subjected to changes and modifications. Some of them can be discarded, because new configurations of the cognitive environmental niche (for example new artifacts) are invented with more powerful offered affordances. Consider, for instance, the case of blackboards. Progressively, teachers and instructors have partly replaced them with new artifacts which exhibit affordances brought about by various tools, for example, slide presentations. In some cases, the affordances of blackboards have been totally re-directed or re-used to more specific purposes. For instance, one may say that a logical theorem is still easier to be explained and understood by using a blackboard, because of its affordances that give a temporal, sequential, and at the same time global perceptual depiction to the matter.

If framed within an evolutionary dimension, the difference between the two opposing views ceases to form a radical objection to the theory of affordance. As just mentioned, we manipulate the environment and thus go beyond the merely instinctual levels adding new or higher-level cognitive skills. The case of stabilized affordances clearly shows that at a neuro-anatomical level a re-distribution between the dorsal and the ventral stream occurs (cf. the last part of section 4.2.2).

4.5 Adapting Affordances and Cognitive Niche Enrichment

In the previous chapter I introduced the notion of cognitive niche enrichment. One of the issues that particular notion is supposed to address is the one concerning *progress* and *innovation*. That is, our cognitive niches may collapse (see section 2.6), but also develop and progress. Cognitive niche enrichment is the phenomenon describing those situations in which a cognitive niche is enhanced. In this section, I will describe the process of cognitive niche enrichment in connection with affordance. More precisely, I will go back to the case of Ambient Intelligence presented in section 3.4 to show how relevant the notion of affordance can be in discussing Ambient Intelligence as an example of cognitive niche enrichment. As it will be clear from the following, my take is that Ambient Intelligence enriches our cognitive niches by providing new forms of what I call "adapting affordances".

4.5.1 Adapting Affordances

Part of the affordances available are delivered by other human beings. Other human beings are so important because they deliver a special kind of affordance, which Ambient Intelligence tries to mimic. These affordances can be called *adapting affordances*. As just stated, affordance detection relies on abductive skills that allow us to infer possible ways to cope with an object or situation. In the case of affordances furnished by other people, the process of detection may be further mediated by social interaction. That is, other people may provide us with additional clues to help us better exploit environmental chances, namely, hidden or latent affordances [Gibson and Pick, 2000].

Care-giving is an example of what I am talking about. As a matter of fact, babies are heavily dependent on other human beings from the very beginning of their existence. They would not survive without constant care administrated by their parents. More precisely, caregivers assist infants in turning the environment into something they can handle by means of learning [Gibson and Pick, 2000]. In doing so they constantly adapt their behavior to create suitable action possibilities that infants can be easily afforded by [Gibson and Pick, 2000]. Caregivers, for instance, contribute to expanding the basic repertoire a newborn is equipped with by manipulating her/his attention. In doing so they prevent the baby from expensive and exhaustive trial and error activities. For example, caregivers:

- act for the infant *embodying motions* that she/he should perform;
- they *show* what the infant is supposed to do;
- they *offer demonstrations* providing gestures so that
- they *direct the infant's attention* to relevant aspects of the environment [Zukow-Goldring and Arbib, 2007].

The influence people have is not limited to care-giving activities. There are a number of situations in which people exhibit the capacity to adaptively alter the representation other fellow humans may have of something. Another example is the so-called *intentional gaze* [Frischen *et al.*, 2007; Frischen *et al.*, 2009]. Recent findings have shown that intentional gaze confers additional properties to a given object, which would not be on display if not looked at [Becchio *et al.*, 2007]. Gaze-following behavior affects the way an object is processed by our motor-cognitive system. More precisely, intentional gaze changes the representation of an object, not merely by shifting attention, but by enriching it with "motor emotive and status components" [Becchio *et al.*, 2007, p. 257].

From a theoretical perspective we may argue that human beings function as a kind of *adapting task-transforming representation*. The term "task-transforming representation" was introduced by Hutchins [1995] to refer to the fact that external artifacts shape the structure of a task – its representation – helping people solve the problem they are facing. A tool may transform the structure of a task:

1. redistributing the cognitive load;
2. rearranging constraints and action possibilities;

3. unearthing additional computational abilities;
4. increasing the number of operations while reducing mental costs.

In the case of *adapting affordances*, the cognitive load is reduced by means of a *transformation* [de Leon, 2002], which adapts the structure/representation of the task to allow a person to detect latent environmental chances. Caregivers and the intentional gaze are fair examples, as they show how people adaptively manipulate the representations their fellows have of the environment to favor or facilitate the exploitation of latent affordances.

4.5.2   Ambient Intelligence and Adapting Affordances

As already argued, Ambient Intelligence can be considered as "cognitive niche enrichment" because of the way it enriches our cognitive niches, that is, by populating them with devices able to keep track of what we do, and then adjust their response adaptively. These various devices are thought to deliver affordances that are somehow *adapting*.

Adapting affordances are those affordances that help the agent exploit latent environmental possibilities providing additional clues. As I have already pointed out, an affordance may be hidden for a number of reasons. For instance, one may lack the proper knowledge required to detect an affordance at the moment of acting. On the other hand, it might remain hidden because of the ambiguity of certain sign configuration so that the agent's function is not immediately intuitive. Finally, affordances may prove unavailable just because a certain person suffers from some impairment – temporary or otherwise – that prevents her from exploiting some particular environmental offerings.

AmI may enrich one's experience of her environment in a number of ways. In the following I am going to present three main cases in which our experience is enriched with respect to three "modalities". *AmI* may eco-cognitively enrich our ability to 1) sense, 2) act, and 3) reason or make decisions. This general classification is indeed arbitrary and it is not meant to cover all the possible ways in which an agent can interact with her environment. However, it may be useful when showing some examples in which "adapting affordances" are provided, and looking at how they enrich the experience of the agent's environment.

The first modality concerns our senses. Sensing something is the basic modality through which we can interact with our environment, and it can basically be related to haptic perception, visual perception, auditory perception, and tactile perception. At times some of our senses may be impaired or less effective for a number of reasons, consequently rendering a person less able to cope with her environment the way she would like to. This happens for example to elderly people, who often suffer from pathologies that impoverish their senses, and thus their wellbeing. Assisted living technologies are meant to give support to those people that have various problems related to aging. An interesting example of Ambient intelligence environment supporting sensing is provided by ALADIN. ALADIN is lighting system designed to assist elderly people in the process of adjusting indoor lighting [Maier and Kempter, 2009].

Generally speaking, light has an important impact on the wellbeing of a person. It affects a number of activities that are indeed fundamental for us. Cognitive performances like reading or concentrating may be drastically affected by light. Light also affects sleep quality, the metabolic system and changes of mood. The impact of light may acquire even greater importance for elderly people, who often have impaired vision or suffer from limited mobility so that they remain at home for most of their time [Maier and Kempter, 2009]. ALADIN is meant to assist them in designing and maintaining indoor lighting.

Basically, ALADIN is capable of acquiring and analyzing information about "the individual and situational differences of the psycho-physiological effects of lighting" [Maier and Kempter, 2009, p. 1201] so as to provide adaptations specific to the user's needs in her domestic environment. More precisely, the system constantly monitors the user's state through biosensors which measure various parameters – elecrodermal, electrocardiographic, respiratory, and those related to body movement. The data are transferred via bluetooth to the computer system and analyzed. Then, according to the adaptive algorithms, the system tries to find the best light modification. In fact, the best light modification is reached by simple feedback provided by the user and from the biofeedbacks acquired through biosensors attached to the user's body.

As already mentioned, lighting may affect a person in a variety of ways, for instance, in the preservation of an active and independent lifestyle or physical and mental fitness [Heschong, 2002]. For instance, elderly people are often unable to correctly assess some internal states with great precision and to then act accordingly. Reduced cognitive performances due to aging may also have a negative impact on their sensing capacity. This, in turn, limits their ability to detect those affordances enabling them to design and maintain suitable indoor lighting to enhancing some activities like reading or relaxing. Biosensors monitor some physiological parameters – as I have just mentioned – and provide some information. This information may help the user modify their domestic light setting in a way that may eventually make them aware of some physiological needs, for instance, the need to rest or have more light in the room. In this sense, the various adjustments the system carries out during a session clearly function like a "mediator" which unearths new environmental possibilities. That is, the system alters the representation of the user's body in a way that allows them to be better afforded by the lights.

The experience of our environment could also be improved, when we act on it. In order to better grasp this point, we refer to the case of the guide sign system. Navigating effectively through certain spaces such as university or hospital buildings could become a problem, sometimes even involving negative emotions like anxiety and tension. Generally, losing the way is annoying and it wastes so much time. Therefore, having guide sign systems that can effectively assist a person to go where they are supposed to go is extremely important.

Moving through a space can be considered as a cognitive task [Conlin, 2009], in which one tries to make sense of the various detectable clues. This is especially true for those spaces that can be labeled as *meaningful*, as they are intentionally designed to bring people to certain places just like in the case of a hospital building.

Meaningful spaces furnish people with a variety of clues, which should help them find the way. For instance, in hospitals the use of the uniform design or the use of same materials and colors. Generally speaking, these clues and guide signs furnish affordances that users can exploit in order to reach the place they are supposed to.

However, the information and clues that guide sign designers can scatter are limited given the various physical as well as cognitive constraints. Physical constraints are given, for instance, by the fact that if space is limited so is the possibility of locating signs. Cognitive constraints are due to the fact that sometimes having more information may cause cognitive overload, and thus waste time. Therefore, affordance detection may become a problem. Ambient intelligence can assist people, for instance, in navigating space providing adapting affordances. Hye described and tested a prototype of an interactive guide sign system meant to help people navigate a hospital, constructing smart environments by means of various devices like *RFID* (Radio Frequency Identification), smart card, and wireless communication [Hye, 2007]. In this case, the use of such devices provide additional affordances that have the main role of helping users exploit latent environmental possibilities, which in this case are basically guide signs.

Adapting affordances can assist people in contexts involving more complex cognitive activities, for instance, reasoning and decision-making. An interesting example about how Ambient Intelligence may help people is illustrated by [Eng *et al.*, 2005]. Using a learning model called "distributed adaptive control" (DAC), Eng and colleagues have designed intelligent environments supporting an interactive entertainment deployed at the Swiss national exhibition Expo.02. DAC was designed to influence the behavior of visitors by learning how to scatter cues so as to make some areas of the exhibition – often avoided or disregarded by the visitors – more visible and easier to reach. Here again, the contribution of the so-called smart environment is to adaptively adjust the system response with respect to what the visitors and users know by furnishing the proper signs.

Another example is provided by the Glass Bottom Float project (*GBF*), which aims to inform beach visitors about water quality so as to increase the pleasure of swimming [Bohlen and Frei, 2009]. Basically, *GBF* is a beach robot monitoring a number of parameters: water and air temperature, relative humidity, pH, conductivity, salinity, dissolved oxygen, etc., but also the speed of boats passing by, people playing on the float, and, more generally, underwater marine life. In addition, these parameters are validated by opinions gathered from interviews with swimmers at the beach.

It is worth noting that *GBF*, coupled with the use of *SPM*, is completely different from the other projects aiming to deliver information about water and air quality. We know that both government agencies and official news releases provide the user with similar data. However, what is interesting in the case of *GBF* is the frequency and quality of the updates it delivers to the public. The way this is done is completely different, and enriches one's experience. For instance, the use of beach visitor opinions potentially increases the reliability of the recommendations. This enables a completely new way of distributing knowledge among beach visitors.

The data gathered are then processed using the so-called "swimming pleasure measure" (*SPM*). Basically, the *SPM* makes predictions about the expected pleasure people may have swimming on a certain beach. The recommendations generated by the system are then delivered in real time to the public via mobile phones. In this way, people can obtain important information before driving to the beach. Here we have an example showing how our ability to formulate judgments and make decisions can be supported and improved by information that would not otherwise be available. More precisely, *SPM* provides us with additional clues that permit us to be *afforded* by the local environment in order to make decisions and carry out reasoning.

4.6   Why and When We Are Not Afforded

In section 4.4 I tried to illustrate how an abductive framework is particularly fruitful in understanding the process of being afforded. The aim of this closing section is, conversely, to survey a number of cases in which people are not afforded, and then to explain why not. The interest in spelling out the reasons why people might be not afforded resides on a very simple claim: a person not being afforded does not necessarily mean that there are no affordances to detect.

I already pointed out that an affordance is a symptomatic configuration of signs informing a person about a way of exploiting the environment, meaning that the environment enables her for cognitive coupling. If this is correct, then we should distinguish between two cases: 1) when the sign configuration is not symptomatic to him/her; 2) when there are no sign configurations at all.

4.6.1   Hidden, Broken, and Failed Affordances

In this subsection I will briefly illustrate several types of affordances: hidden affordances, broken affordances, and failed affordances. Let me start with the first type. A person might not be afforded because she cannot make use of certain signs. Basically, this happens because he/she lacks the proper knowledge in terms of abductive skills suitable for interpreting the signs or clues available in the environment. This is the case of *hidden* affordances previously introduced in section 4.2.1.

One of the simplest examples of hidden affordances regards babies. Babies are usually not afforded by their parents' environment, simply because they are still developing the repertoire of skills necessary for detecting even simple affordances. For example, infants of age 8-12 months are not yet able to detect the affordances of a spoon for eating [Gibson and Pick, 2000]. They simply bang or wave the spoons they are given. Indeed, using a spoon for eating requires a number of perception-action skills that the infants are still developing.

From our perspective, the infants are not yet able to detect some affordances because their repertoire of abductive skills – pre-wired by evolution – requires a period of training. That means that the brain structures underlying certain abductive skills, even though already pre-determined at genetic level, have to be further specified by the relevant experiences the infant has in an enriched environment [Marcus, 2004].

Another interesting case in which a person might be not afforded is provided by so called *broken* affordances.[9] Broken affordances regard all those people suffering from brain damage or some other injury. The impairment may be temporary or permanent, either way they simply miss some affordances just because they lost some crucial neural mechanism underlying those skills required to detect the affordances, which otherwise would be immediately available to them. For instance, patients affected by *optic ataxia* (dorsal stream impaired) are able to name an object appropriately and recognize its function, but remain unable to grasp and locate it to exploit its affordances. As already mentioned in section 4.2.2, what they lack is the ability to unconsciously adjust ongoing movements something that seems to suggest the existence of an automatic pilot, which in this case is impaired [Himmelbach *et al.*, 2006, p. 2750]. In this case, damage to the dorsal stream prevents the patient from making use of a number of wired abductive skills – mainly manipulative and kinesthetic – which would have allowed him to detect the affordances available.

An affordance might not be detected because it is poorly designed so that a person can hardly make use of it, meaning that the sign configuration is poorly symptomatic. In this case we have what I call *failed* affordances. As already mentioned, Norman introduced the distinction between *potential* affordances and *perceived* affordances precisely because he wanted to draw a line of demarcation between merely potential chances and those that are actually exploited by the user [Norman, 1999a]. He argued that only in the second case we do have affordances. However, it is worth noting that it is nearly impossible to predict whether or not a user will detect or perceive the affordances constructed by the designers.

Indeed, the cooperation between designers and users can be enhanced so as to allow designers, for instance, to characterize and thus predict the most likely user reaction. However, like any other activity involving a complex communicative act, there are always design trade-offs [Sutcliffe, 2003] between user needs and environment constraints identifying not the optimal solution, but the satisfying one. Design trade-offs simply indicate that some affordances intended by the designer cannot be optimally constructed, meaning that an affordance, like a sign configuration, might be ambiguous to the user. If so, then the distinction between potential affordances and perceived affordances starts blurring. Let me see how my conception of affordance may be helpful to solve this problem.

As already pointed out, an affordance informs us about environmental symptomaticity. That is, an affordance is a sign configuration that is totally or scarcely ambiguous so that we promptly infer that an environmental chance is available to us. Therefore, the ambiguity of a sign configuration can be a valuable indicator to demarcate the boundaries of what can be called an affordance and what cannot. However, it is worth noting that on some occasions the user fails to detect an affordance simply because of design trade-offs. For we propose to use the term failed affordances to indicate those situations in which a sign's configuration might

---

[9] I derive this expression from Buccino *et al.* [2009].

favor some misunderstanding between designers and users as resulting from design trade-offs. Notwithstanding the fact that they are ambiguous – to a certain extent, failed affordances are still affordances, as their ambiguity is more a result of design trade-offs than of the absence of symptomaticity. Failed affordances are common, for instance, in HCI when designers simply fail to build a mediating structure – an interface – able to communicate the intended use of an object.

4.6.2   Not Evolved and Not Created Affordances

In this last subsection I will consider affordances with relation to creativity and evolution. My main claim is that a person may not be afforded by the local environment in a certain way, just because there is nothing to be afforded by. This may happen for two reasons. First of all, because a configuration of signs supporting a certain activity has not been discovered yet. Secondly, because that configuration of signs has not yet been selected by evolution; and maybe it never will be.

As regards the first category, the fact that there is nothing to be afforded by simply means that our abilities as eco-cognitive engineers are limited. As already mentioned in the first section, the limits of the agent's adaptability correspond to his limits in creating affordances. I have previously discussed failed affordances arguing that the detection of environmental symptomaticities may be partly impaired by ambiguity due to some design trade-offs. The ambiguity of a sign configuration may lead us to misunderstand how to exploit our local environment. It may take us more time to detect some hidden chances, but that does not mean that we are totally blind about them. By contrast, sometimes we simply lack any environmental symptomaticities to make use of, because they are not present (yet).

More generally, I may view the creation of new affordances as part of the history of technology which humans are involved in as eco-cognitive engineers. Indeed, the history of technology is a very interesting field as it offers many examples concerning affordance innovation and creation. An interesting example is provided by the so-called *paperless world*. The digital revolution has drawn our attention to fascinating scenarios in which digital technologies – like PCs or tablets – would make paper documents useless or just something from the past. Actually, the history of what we might call "the quest for the paperless world" is extremely interesting when noting the amazing creativity of humans, but also some present limitations [Sellen and Harper, 2002]. The advent of a paperless world has partly come about, as digital technologies have created new affordances. Basically, designers (and users) have unearthed chances for new interactions, in my terminology, new symptomaticities. For instance, the paperless office has provided new affordances related to archiving, browsing and searching through documents, etc.. Designers have been able to make a screen touchable enabling users to underline and manage their notes in a completely new way.

However, some affordances are still missing or, at least, some affordances cannot be reproduced in the digital. An interesting case worth mentioning here is provided by post-it notes. Post-it notes are crucial in everyone's life for managing personal information. More precisely, post-its help us store "information scraps"

[Bernstein *et al.*, 2008]. Information scraps allow a person to hold a variety of personal information like for instance ideas, sketches, phone numbers and reminders, like to-do lists. Generally speaking, information scraps contain all the personal information that is too unexpected or miscellaneous to be encoded or stored in other forms than scribbled notes.

However simple they may seem, post-its are amazingly useful as they contain a number of affordances that allow us to have access to a kind of information that otherwise would be completely out of reach. For instance, a post-it is as ready-to-use as a corner of a sheet of paper. But unlike the corner it is 1) portable, as it can be put in a pocket, 2) it can be stuck on any other object like a computer monitor, a door, a window, etc., or 3) it can be handed to somebody else.

Although there are a number of computer programs that have been appropriately designed to help people take quick notes, the conclusion reached by Bernstein *et al.* [2008] is that at the moment there are no computer applications able to surpass post-its. By definition information scraps are taken down on the fly, and usually a computer program – whatever it is – it takes too long. For instance, even assuming that we have our laptop already working and available to take a note, it forces us, to make a type assignment or assign a category in order to make our note easily retrievable [Bernstein *et al.*, 2008]. But this is precisely what we do not need to do when using a post-it. We do not need to think about how to categorize a certain piece of information or about how to set a deadline. We simply assign a particular meaning and a particular place to a particular pad of post-it so that it will be easily accessible and retrievable later on. Plain and simple, no computer applications afford us the way post-its do. In my terminology, designers have not yet figured out how to embed in the digital those symptomaticies signaling the affordances of a post-it. Here again the limits of being afforded are the limits of human beings as eco-cognitive engineers.

As already mentioned, there is an additional reason why we are not afforded, that is, when an affordance or a set of affordances have not been secured by evolution. In section 4.4 I argued that some affordances do not seem to require any mediation of knowledge, because they are already made available by evolution in terms of pre-wired and thus neurally pre-specified responses to the environment. Let me make a very simple example of affordances that are somehow secured by evolution, the case of the *opposable thumb*. In this case, some hominids developed a particular adaptation enabling them to oppose the thumb to the palmar side of the forefinger. That was a fundamental step in evolution: the opposable thumb made available a number of affordances that were simply unavailable before, like, for instance, grasping, handing, throwing, and so on. Those species that had not develop the opposable thumb simply lack certain affordances. For them, a tree branch is not graspable, they cannot pick something up and then throw it to hit an animal, and they cannot climb a tree.

Another example is that one of the human-chimpanzee common ancestors, the so-called *Ardipithecus*, had a *grasping foot*. However the hominid foot then evolved so that our later ancestors lost it. Indeed, that loss made a number of affordances unavailable, like those related, for example, with climbing trees or locomotion. In fact we still miss them and we can only use our feet for walking or standing, but not for grasping. This has a very simple consequence in that all the objects around us are designed bearing in mind that we do not push a button or type using our feet and we would simply miss such affordances, because they have been put out of reach by evolution.

More generally, what the two examples point to is that evolution makes an organism develop different affordances. That is, there are certain sign configurations that become meaningful for one organism, but not for others: evolution operates in such a way as to allow certain organisms not to be afforded by certain things. In this case, an achievement (that we can consider an adaptation) was not established, simply because it was not profitable for the organism.

4.7 Concluding Remarks

In this chapter I have shown how the notion of abduction can account for some of the ambiguities of affordance. The main thesis put forward here is that affordance concerns all those situations in which the signs and clues we can detect prompt or suggest a certain action. In this way, a chair affords sitting in the sense that clues are available from which we can infer (in the Pearcian sense) that we can sit down on it. The process through which we perceive affordance is entirely abductive.

Humans can also create, modify, and – on some occasions – stabilize affordances that, in turn, become a part of the eco-cognitive inheritance, which humans can take advantage of in order to accomplish various tasks and activities. For ontogenetic processes like social and individual learning play a key role in developing new ways of interacting with the environment and better direct our adaptation. In doing so, humans are involved in a continuous process of crafting and modifying ecological niches aimed at developing new sets of affordances.

Indeed, some major issues still remain open to debate and need further development. As argued in the second part of the chapter, I have tried to stress the evolutionary dimension of affordance, often discarded by researchers and scholars. The thesis I have discussed is that some affordances are somehow pre-wired, whereas others can be created and become stabilized over time. This particular issue can be split in two. The first part concerns integration between the ventral and dorsal stream. How do they interact? Do familiarity and experience re-assign and re-distribute functions to the two systems? As outlined, no conclusive evidence has been reported on that matter. The in the second part we should more carefully detail how various eco-cognitive inheritance structures contribute to the stability and sharing of new affordances. I briefly referred to a variety of means by which affordances are stabilized, they could be behavioral or symbolic, for instance. Recent studies on niche construction may give valuable hints about how to develop this particular issue respecting a broader theory of cognition and its evolution.

Chapter 5
The Notion of Docility: The Social Dimension of Distributing Cognition

Introduction

In section 2.5, I argued that the mechanism for de-biasing rationality is not furnished at an individual but at a *eco-cognitive* level. As a matter of fact, a great amount of knowledge and information is stored in cognitive niches in the form of artifacts, institutions, habits and so on. The question is, which *socio-cognitive* attitudes exhibited by the single human being favor the distribution of cognition and thus the de-biasing of rationality? This chapter will introduce the notion of docility as the attitude supporting and facilitating *cognitive cooperation*.

In section 5.1, I will start by claiming that the traditional notion of altruism offers a poor explanation of cognitive cooperation. The main idea I will discuss is that altruism is not integrated into any cognitive framework potentially explaining 1) why human beings tend to share cognitive resources with other people, and – most of all – 2) why this turns out to be key to the astonishing success of human beings as eco-cognitive engineers.

In section 5.2, I will illustrate the notion of *docility*. First introduced by Herbert Simon in the early Nineties, docility is thought to be the concept able to connect a pro-social attitude like altruism to human cognition. Docility will be described as the attitude or tendency underlying those learning processes, which involve various forms of reliance on social channels.

In section 5.3, I will develop the original notion elaborated by Simon bringing it into the broader framework of cognitive niche construction. As an adaptive consequence of cognitive niche enrichment due to living in large group, docility is a key component in explaining the tremendous success humans had as eco-cognitive engineers. Docility will be defined as that disposition meant to facilitate the exploitation of various eco-cognitive resources in two senses. First, it is manifest in the tendency to lean on social channels in order to overcome individual limitations. It facilitates the exploitation of second-hand information. Secondly, it is not only involved in merely passive attitudes related to obtaining resources, but also in those activities in which one is actively engaged in providing and sharing them.

E. Bardone: Seeking Chances, COSMOS 13, pp. 101–123, 2011.
springerlink.com

After the illustration of docility, I will devote section 5.4 to developing an account of undocility. Little attention has been placed on the definition of this concept, thinking that undocile behaviors are simply those behaviors that cannot be defined as docile. In order to fill this gap, I will resort to the concept of bullshit, first introduced by Harry Frankfurt in his work *On Bullshit*. Taking advantage of this concept, I will argue that Frankfurt's definition of bullshitting captures an important aspect characterizing undocile behaviors: bullshitters, lacking docility, do not display any concern about what they believe. That is, they are basically careless about the truth-value of what they say or maintain. After introducing this interpretation of undocility as bullshitting, I will complete the picture of undocility by pointing to some situations in which docility has an important limitation. In this part of the chapter, I will introduce and deal with the so-called "ostrich effect" in order to show how silence and active avoidance – although undocile – may turn to be beneficial for the individual and the community she lives in.

In the last section 5.5, I will present the Open Source Model as a case in point for better describing the relevant features of docility. More precisely, I will point out that *hackers* – those who are actively involved in the various open-source communities – best represent what a docile person is.

5.1 Altruism and Social Complexity

As a matter of fact, the notion of altruism assumes a great variety of meanings depending on the context in which it is actually considered [Khalil, 2004]. For instance, it is altruistic to help another human being who suffers or is in danger for some reason. This is a kind of *pro-social behavior*. Altruism also means to sacrifice one's fitness in a way that advantages another individual's fitness [Simon, 1993]. Or it can be viewed simply as the intention of benefiting another individual: in this case altruism is a kind of genuine regard for others [De George, 1999].

As the reader can see, these meanings overlap in part, since all of them refer to a certain person who is motivated to give something, but gains (or thinks to gain) no return from the beneficiary of his/her action [Khalil, 2004; Simon, 1990; Simon, 1993]. Roughly speaking, altruism can be described as a *pull-push* mechanism. That is, altruism (and its contrary, namely, selfishness) deals with all those situations in which a person decides to *push or pull* towards oneself or others a certain good (money, happiness, help, fitness, etc.). If one pulls a certain good towards oneself she/he is commonly described as selfish, whereas if one decides to push it towards others, she/he is described as altruistic. The concept of altruism has been commonly used to explain, for instance, *cooperation* [Axelrod, 1984], and all those behaviors that make social living possible. But it is also referred to as a pro-moral behavior that can also be found among animals and insects [Joyce, 2006].

Although I do not find any fault in this characterization, I maintain that this representation is incomplete, since it reduces an interaction between human beings basically to *an exchange of something*: it does not address any *cognitive* dimension. That is, interacting with other human beings can constitute the basis for accomplishing various tasks and, ultimately, for making decisions and solving problems.

The social dimension turns out to be a valuable resource [Frank, 1988; Frank, 2004] that becomes part of our cognitive system and the way it evolves [Humphrey, 1976; Magnani, 2007c].

As it will be shown below, my perspective tries to enhance these perspectives since we found that cognitive dimensions provide a better understanding of human nature rather than the concept of altruism. However, I acknowledge that altruism provides a consistent example of a pro-social human trait. Altruism is not a simple selfless act and clearly goes beyond ego fulfillment. Having written that, it is apparent that I to some degree follow Khalil's altercentric perspective since the individual "at least in some occasions, may share income because he is built with a pro-social trait" [Khalil, 2004].

An enormous amount of information is stored in human culture and becomes available only through social learning. Information is also stored in institutions, language, and other various artifacts we daily cope with (computers, books, etc.). Human beings are *ultra social* creatures [Richerson and Boyd, 2005], because living in large groups becomes a fundamental trait that contributes to shedding light on some human activities, for instance, decision-making and problem-solving. On this tack, Dunbar [1996; 1998] extensively investigated the impact that social complexity – typical of human communities – has had on our brains. According to the so-called "social brain hypothesis", there was a co-evolution between (a) the human brain and (b) living in larger and larger groups during the history of human evolution. Human beings developed disproportionately those skills which allowed them to better cope with the social complexity and the increasing cognitive demands related to living in larger group size.

Living in larger groups gave birth to a series of events, which had clearly a major impact on human evolution, because it drastically reconfigured the attunement and access to those relevant aspects of the environment, which enabled human beings to exhibit more and more plastic behaviors. In this sense, the revolutionary character of social complexity in the history of human evolution is closely related to its eco-cognitive effects at the niche-construction level. That is, social complexity profoundly changed and enriched the eco-cognitive niches our ancestors lived in, and so contributed to that cognitive explosion that eventually led to their ecological domination as self-domesticated animals.

The idea of humans as *self-domesticated* animals is introduced by Bingham [1999] He connected this idea to the formation of early moral communities, which played a pivotal role in controlling and modifying various pre-existing instincts towards the social and cooperative living. Recent comparative studies of monogamous species [Curley and Keverne, 2005] reported reduced constraints of hormonal states over parenting behavior even in Old World Primates: that is, parenting behavior is much less influenced by hormonal states, but other brain-based mechanisms. From the evolutionary history, this resulted as an adaptation to living in larger groups. Significantly, the decline of the mechanism underlying olfactory processing is supposed to have been compensated by the emergence of an alternative brain-based mechanism based on foraging information from visual clues. Therefore, our ancestors started relying on different mechanisms much more devoted to making use of external and visual signs. One of the most intriguing differences is, for example, the development of various social instincts, which enabled the cooperation also among individuals not genetically related. Worth considering it is also the evolution of social skills related to extra-species relationships with domesticated animals. As argued by Adolphs [2001], a complete treatment of social human processes should include reciprocal social behaviors with those animals which entered human niches.

The general reconfiguration of the relevant aspects of the environment, namely, niche construction activity (cf. section 3.3), had some important adaptive consequences; For instance, the evolution of *special* cognitive processes to cope with social complexity. According to recent neurological studies, for instance, social cognition is special or at least involves neural structures that organisms can only modulate for other limited cognitive performances, like decision-making or attention [Adolphs, 2006]. Usually people think of other humans in normative terms, whereas thinking about inanimate objects does not have any normative character. As a matter of fact, we are not used to gossip about a tree or even an animal, because we do not think about them in normative terms. The same can be said about *mind-reading*. People surely attribute human features to inanimate objects (even moral instrinsic value), but rarely they have an object as a mind-reading target [Adolphs, 2006].

More generally, human beings developed disproportionately those brain areas underlying the skills, which allowed our ancestors to better cope with the increased cognitive demands derived from living in large groups. This is fairly captured by the so-called "social brain hypothesis" [Dunbar, 1996; Dunbar, 1998]. For instance, the ability for identifying individuals and their behavior through visual signals detection; the ability for face recognition; the ability for remembering relationships in a group and manipulate them; the ability for acting on emotional cues given by the others; the ability of mind-reading, etc.

Thus, primate brains would be biased towards social problem-solving since their evolutionary origins [Humphrey, 1976; Byrne and Whiten, 1997] in the sense the eco-cognitive relevant aspects of the environment – their niches – humans are mostly attuned to are *social* in their origins. As shown above, language – in its gossip-enabling features – was a tremendous adaptation for social exchange. In this sense, language allowed humans to develop new and powerful cognitive niches resulting from the emergence of social complexity given its artifactual nature. Language as the ultimate artifact [Clark, 2006; Clark, 2008] contributed to form a permanent artificial environment to support an astonishing variety of activities [Magnani, 2009]. Also the gossiping fallacies can be easily considered in this distributed and artifactual framework (cf. Chapter 1).

It is noteworthy that what originally developed to compete within the social *melieu* shades off into other extra-social domains. For instance, the ubiquitous propensity of human beings to anthropomorphize even in presence of stimuli that are not strictly social. Interestingly, the projection of some human features to various external objects is less trivial than one might expect. Anthropomorphic thinking, for instance, is of that importance when dealing with moralizing various objects [Gebhard *et al.*, 2004]. The recourse to projection can be useful for unearthing new interpretations and points of view, which were previously unavailable.

Even if anthropomorphic thinking is a kind of pre-scientific thinking, it has that heuristic value in many human activities just like the case of gossiping fallacies I treated in Chapter 1, which are, indeed, fair examples of what we are talking about. Another interesting example is provided by the so-called "social illusion" originally studied by several scientists since the second half of the last century. That is, simple triangles or circles moving a certain way may trigger emotions and social description [Castelli *et al.*, 2000; Adolphs, 2006]. Surprisingly, those projections recruit the same visual cortical regions involved in face recognition [Schultz *et al.*, 2003]. Whereas patients affected by autism are unable to socially describe or even have an emotional response while viewing those shapes [Adolphs, 2006].

5.2 From Altruism to Docility

The concept of altruism accounts for the social dimension of ultrasocial creatures like us, but it is quite lacking in its cognitive component, which is clearly fundamental, as the evolutionary studies I mentioned clearly point out. What is lacking in the very notion of altruism is that it does not provide any hints about how the social (and cognitive resources embedded in it) contributes to shape the way we make decisions and solve problems. What Simon [1993] did when he introduced the notion of *docility* was to bridge this theoretical gap.

As a matter of fact, most of the time people rely on information and suggestions they gather from others. Before buying a laptop, for example, we usually consult a friend who is thought to be competent in the matter (if we have one). We simply trust his suggestions and we subsequently disregard other kinds of information, for instance the technical advice that a sales assistant gives us. The kind of relationship we have with a given person turns out to be a valuable cognitive resource. This is the case, for example, of group loyalty as a form of altruism. Simon argued that group loyalties have not only a motivational dimension, but also a cognitive one. He argued that "[t]hey define the boundaries of the group over which goods are to be summed, and they cause particular variables and simplified world models to govern the thinking of group members" [Simon, 1993, pp. 159-160]. Indeed, there are plenty of examples of that kind. Now, the point that Simon makes is that being open to others, often labeled as "altruistic behavior", turns out to be a fundamental trait (or feature or strategy) related to how humans face difficulties and overcome their various cognitive limitations to make satisficing decisions.

Simon calls this tendency to lean on social channels *docility*. Humans are "docile", in the sense that *their fitness* is enhanced by "the tendency to depend on suggestions, recommendations, persuasion, and information obtained through social channels as a major basis for choice" [Simon, 1993, p. 156]. Humans support their decision-making capabilities by receiving inputs, perceptions, data, and so on, from the social environment. The social context gives them the main data filter available to increase individual fitness. As a matter of fact, everything we know does not derive from our own experience, but is learnt from those who surround us [Humphrey, 1976; Magnani, 2006a]. Very often we simply watch what others are doing and imitate them. Imitation is indeed one of the most common learning activities that we may find in everyday life. Just imagine how our life would be if we had to experience everything by ourselves. Then, I argue that the environment and other people's attitudes make the difference in developing such a tendency. In this sense, the fitness of docile individuals depends on other people willing to provide advice, comments, suggestions, and the like. Moreover, the social system leads docile individuals to a better fit if compared to non-docile ones.

Although Simon never addressed any of these "distributed" conceptions of rationality or human cognition, as I have already described in Chapter 1, it seems reasonable to say that the notion of docility clearly introduces a new element to the theory of bounded rationality. Simon's statement can be fruitfully interpreted that way: humans overcame the limits of their bounded cognitive system by *delegating* cognitive functions to the environment. Suggestions, recommendations, and the like are all external resources that are socially available, and that indeed contribute to lessening various limitations. They can provide both short-term and long-term aids to reduce memory load. Within scientific and academic settings, for example, a sound comment from a reviewer can help us to direct my attention to a point we overlooked. An experienced colleague of ours can provide us with knowledge and skills which we can easily adopt upon occasion, for instance, to organize a conference or write a paper. A discussion may unearth new perspectives that were not previously available within our minds. It can be also a very precise system for getting feedback to test some ideas we are not sure about. And so on.

Other people provide external resources, suggestions and recommendations, but so do the various objects and tools we are familiar with. For instance, a pen and a sheet of paper are an example of a simple but powerful external cognitive kit that drastically increases our performance when dealing with calculating, planning, remembering, communicating, thinking and other related activities. More generally, artifacts:

1. support memory [Wilson, 2005];
2. change the representational task we face in terms of more efficient action sequences [Zhang and Patel, 2006];
3. facilitate the making of inferences which are more sound than others.

Going beyond Simon, docility can be considered as a kind of *adaptation* that facilitates the process of distributing cognitive functions to the environment, and makes that a major basis for decision making. From our birth we operate this kind of delegation, first to our parents, and then to other people. After that we begin to select and distinguish between people from whom to *learn* something important or insignificant, and the importance of a personal role in getting information becomes ever greater. This is also apparent in studies on advice-taking and -giving where scholars found that people discount information according to whether it comes from expert or novice advisors.

5.3 Docility, Learning, and Knowledge

5.3.1 Developing Docility: The Active Side

A great part of what we know is derived from other human beings and, more generally, from social channels. Just looking around, we can see a great variety of systems that make information transmission possible and let it work as a major basis for choice. Now, I want (and need, as shown above) to connect docility to cognition on a broader but definite basis. Hence, I argue that docility is that tendency which humans exhibit in any cognitive activity that mostly leans on what is found outside – in the human cognitive niches.

As repeatedly pointed out, the human cognitive system is distributed so that some performance results as a continuous interplay between our mind-brain system and various external resources [Clark and Chalmers, 1998]. That is, people do not hold internally an explicit and complete representation of a given task that they face and its variables (cf. section 3.1). Rather, people continuously adjust and refine their perspective through a further exploration of the environment, which allows them to get a more detailed understanding of what they are doing [Thomas, 1999]. Various resources (both animate and inanimate) are picked up upon occasion, and external counterparts, namely external representations, are created to enhance the quality of their performance.

Having assumed this updated perspective on human cognition, Simon's definition of docility is quite lacking, since it postulates docility as just a passive attitude. Conversely, I claim that it also has an active side because it deals with the entire process of distributing cognitive functions in which a person actively exploits external resources that are socially available (cf. "the externalization process" discussed in section 3.1.2).

Simon argued that some people are docile in the sense that they lean on suggestions and recommendations by others. But I can extend that definition also attributing docile behaviors to those who *provide* suggestions and recommendations. In this case, people are docile because they tend to share with others what they know, namely, suggestions, information, etc. In doing so, they contribute to creating and generating a common basis for communicating and solving problems. Thus I modify Simon's definition as follows: docility is the tendency to depend on suggestions, perceptions, comments, and to gather information from other individuals on the one hand, and to "provide" information on the other.

This definition may contribute to the development of a new model of social interaction which is worth investigating. According to Simon's definition, docile people are those who lean on external supports like suggestions, comments, and so forth. As he noted, we are all, to some extent, docile because we live in society, and that cannot be discarded. However, Simon's perspective turns out to be oversimplified as well. If humans are basically social beings, the degree to which they contribute to society may vary drastically. Therefore, active and passive docility cannot be found the same way in each individual.

The new definition I provided may bridge the above mentioned theoretical gap. As I put it, docility has both an active and passive side. Developing this line of thought, the active side can be further articulated into three main elements; thus, docility can be viewed also as the tendency

1. to share one's own information;
2. to give a public and social dimension to one's thought/work;
3. to render communication easier by creating, maintaining, and developing standards, or standard-fidelity.

Information sharing. This is recognized as one of the main attitudes needed in organizational and social life [Humphrey, 1976], and it may also be characterized as one of the traits of human personality. Differences between more active and more passive docile individuals depend on the quality of the information shared. For example, we should write that people with that particular tendency to share information on values, assumptions, beliefs, and expectations show clear attitudes towards leadership. Of course, I suggest that this makes the difference between active and passive docility attitudes.

> Information sharing makes the difference between docile and non-docile attitudes in individual behaviors.

> The quality of the information shared makes the difference between more or less docile individuals. The more the quality, the more the individual is suited to fit in the social context.

General interpretation of the model indicates a tendency for differences between individuals such that we cannot find a completely non-docile or a fully docile individual. Thus, the sharing of information is related to the tendency to behave more or less cooperatively, i.e. the more the individual shares information, the more he or she shows a docile attitude.

The Public and social dimension. I maintain that docile people think socially [Kunda, 1999]. As already noted, living in large groups of people drastically contributed to survival and reproduction [Dunbar, 1996]. As a matter of fact, the larger the group size, the more it was possible for group members to protect themselves from danger. Indeed, this does come at a cost, because it implies various bounding mechanisms to make collective decisions and manage social complexity. Within this framework, docile people are those who tend to build communities of practice and learning as the basis for sound decision making.

> Docile individuals need to build communities.

Of course, trust and commitment play an important role in this process. However, people behave differently as communities evolve, and we argue that active docile individuals tend to build communities while passive docile individuals tend to lean on these communities. As a result, I argue that:

> Docile individuals who build communities know better than others how to exploit social channels.

Standard-fidelity. Docile people exhibit the tendency to share information, ideas, etc. That means they are committed to rendering their communication as clearly as they can. Indeed, if one wants to share an idea, it is obvious that he/she must try to make it plain and clear, otherwise he/she would fail in his/her goal of passing on information. That is to say, docile people tend to follow standards. This tendency, which can be called standard-fidelity, has important cognitive relevance, because it makes information and knowledge transmission much easier. Consider, for instance, the case of mathematicians, logicians, etc.: as much as possible, they employ ways of communicating and transferring knowledge that are transparent, namely, standardizing, to overcome the ambivalent character of natural language. Thus, mathematicians and logicians have to follow the rules and standardized procedures embedded in the channel (or mediator) they use to communicate [Magnani, 2007c]. Now, it is worth noting that docile people exhibit a tendency to share and use standards, but they are also often involved in redefining them to enhance the transmission of knowledge and to make it more efficient. Since docile people are well-committed to passing on information and distributing cognitive functions to external resources, they focus more attention than others on the social channels, as Simon talked about. However, I argue that there is a difference between people who create and develop standards, and people who use and maintain these standards. From this perspective I may define non-docile individuals not using standards at all, or using them improperly.

> Creating and developing standards characterizes higher degrees of docility (active more than passive).

> Using and maintaining standards characterizes lower degrees of docility (passive more than active).

Indeed, it is not possible to have pure types of docile people. For instance, pupils can be passively docile when listening to a teacher's lesson, but they are just as docile when they actively participate in the learning process, for instance, when they ask questions or attempt to answer them. In these cases, they cease to be passive and provide information that a teacher can fruitfully take advantage of.

Standard-fidelity is one of the most interesting aspects concerning docility. In order to introduce this notion, let me make an example. Consider the difference between a mathematical theorem and a magic trick; mathematicians and magicians simply differ in the method or the procedure they carry out in order to get the result. A mathematician has to follow rules and procedures that are accepted as objective. For instance, one cannot use theorems that have not yet been clearly demonstrated. Any passage must be justified according to the laws of logic: neither contradiction nor partiality can be accepted. In contrast to that, a magic trick is something completely private. That is to say: first of all, it is not publicly available to everyone who wants to know anything about it; secondly, the procedure through which one can make the trick work is kept secret as well, known only to those within the magic circle; third, there is no standard at all, since any magician can perform tricks on his or her own.

Generally speaking, I may say that the mathematician has to follow certain standards that are the standards of mathematics, accepted within the field. There is some kind of practice, such as proving a theorem or making a scientific experiment, that requires people to follow how (a) certain resources that have been employed by others and (b) the rules embedded in them. In this case, docility is represented by what I call standard-fidelity.

What is the main reason why docile people are supposed to use and share standards? By definition, docile people are those who are committed to share and exploit social channels as a way of solving problems and making decisions. Now, the connection to standard-fidelity is that using standards makes information i) transmission, ii) exploitation, and iii) re-use much easier. In this sense, standard-fidelity explicitly concerns the increasing of standards as a major opportunity to accomplish these three main docility-dependent activities. Consider a scientific experiment: here scientists follow certain standardized procedures that clearly display results and the way to test their presumed validity. That is, standardized procedures lead to results that can be understood more easily and shared better by the community of practitioners (scientists, mathematicians, and so on).

The analysis of standard-fidelity implies making explicit some basic assumptions which are not part of the definition of docility, but which are prerequisites - so to speak. First of all, standard-fidelity assumes that the actors at play posses the skill and ability to achieve clarity as a basic requirement for communicating ideas and thoughts. This is connected to the idea of using language as a public tool normally lacking in people suffering various pathologies related to autism, for instance. Besides that, I also assume that people posses the ability to approve or disapprove of a certain behavior with relation to a rule or a set of rules. Standards can be easily deemed as surrogates of those behaviors which are concerning with approving or disapproving certain acquired ideas. This role is basically played by morality in human societies [Castro *et al.*, 2004].

In the next section, I will try to expand my framework on docility trying to account for the intrinsic connection to the process of learning. As Bruderer and Singh pointed out, docility relates to "specific learning capabilities" [Bruderer and Singh, 1996, p. 1324], and I may add that this is in line with the Latin origin of this word (late 15th Century). Docility means "apt or willing to learn", from Latin "docilis", deriving from "docere", i.e. to teach.

5.3.2 Docility, Learning, and Competence-Dependent Information

By definition docility is involved in any process of acquiring additional resources or knowledge by means of a disposition or tendency to learn from social channels. To work out this definition and its implications, consider the *ignorance problem* I illustrated in Chapter 1. As already mentioned, we are facing an ignorance problem when we have a cognitive target T which can be attained by a piece of knowledge K that we lack at a certain time t. As noted an ignorance problem can never be solved, but just *transformed* into something manageable. The price paid is that our commitment can only attain a lesser target.

In that section, I described two transformative options to the ignorance problem. The first resorts to generating hypothesis H which, if true, would hit T. This option is suggested by Gabbay and Woods. The second option resorts to the resources one can exhibit as part of a social group. For instance, the three fallacies I described in Chapter 1, are examples of this second transformative option.

Taking the notion of docility described so far, the two alternative options I just mentioned differ as regards another important point, that is, whether they lean on docility or not. I will rename the two options calling them respectively the *docile* option and the *non-docile* option. Specifically, the docile option is the one in which we try to attain T by means of attempting to learn K – or conjecturing it. The non-docile option is when we try to attain T attempting not to learn K. These two definitions clearly point to learning as the major cognitive asset involved in docile behavior. This statement can be further clarified, when connected to the notion of competence-dependent information, that I introduced in section 2.4.

This seems to suggest that our capacity to make use of knowledge, which ultimately leans on developing the appropriate abductive skills (cf. 2.5.1), cannot be dealt with without referring to docility. In fact, using competence-dependent information clearly requires a learning process to happen and docility would be the disposition underlying such a choice. In this sense, docility can be considered as a truly Socratic attitude, because it involves becoming aware of our ignorance, and acting consequently to try to overcome it by means of learning.

Conversely, the non-docile solution does not require us to be involved in any learning process. That is to say, we employ the competence-independent information, which allows us to reach a decision, without resorting to any knowledge acquisition. This solves our ignorance problem in an *anti-Socratic* manner keeping us in an *ignorance of ignorance* status. The price we pay for economizing is failing to become aware of our ignorance.

I are now in the position to argue that docility is more generally related to the tendency to lean on the various *ecological* resources, which are released through the cognitive niches. Combining that argued so far with the evolutionary approach described in Chapter 3, I posit that docility is:

- a disposition underlying those activities of externalizing cognitive functions, which are connected to the delegation and exploitation of ecological resources;
- a disposition that permits the inheritance of a large amount of useful knowledge while avoiding the costs related to (individual) learning;[1]

---

[1] The notion of docility might resemble that of social learning. The two notions are indeed closely related, but different. Social learning describes the way acquired traits can be learnt from other people, whereas docility is more related to decision-making, that is, it is more closely related to exploiting various ecological and social resources not to learn, but to make decisions.

- an adaptive response to (or consequence of) the increasing cognitive demand (or selective pressure) on those information-gaining ontogenetic processes, resulting from intensive niche construction activity.

Indeed, all eco-cognitive resources do not simply come from other human beings. This is clearly an oversimplification. However, the information and resources we continuously exploit are – so to speak – *human-readable*. Both information production and transfer are dependent on various *mediating structures* [Hutchins, 1995], which are the result of more or less powerful cognitive delegations, namely, niche construction activities. Of course, it is hard to develop and articulate a rich culture, as humans have, without taking into account the role of mediating systems.

5.4   Who Is Undocile?

5.4.1   Bullshitting and Undocility

The notion of bullshit introduced by Frankfurt [2005] may help to describe a fundamental feature of people who are not docile, this relates to their carelessness about the truth.[2] According to Frankfurt, there is an important distinction to draw between a bullshitter and a liar.[3] What differs between the two is that the liar has a general concern about truth. This is true because, in order to tell a lie, he has to know – at least – what the truth is. Although the liar fails to be cooperative with respect to the content of a certain state of the world, he is indeed cooperative with respect to his attitude towards truth. The fabrication of a lie may require a great deal of knowledge and it is mindful as it requires the guidance of truth.

More generally, a certain state of mind – namely, an intention to deceive – is assumed by the liar while making a statement. This attitude is what makes his statement potentially informative. For instance, consider the case of a person telling us that he has money in his pocket, when he has not. His lie is informative as one can guess whether he lied or not. What is interesting about lying is that there is always a reason why a person may not be telling the truth, in fact lies and deceit can be detected. People, for instance, have at their disposal both verbal and non verbal cues enabling them to detect potentially deceiving situations [Vrij, 2008]. A minor detail of dress may suggest to a man that his wife is cheating on him, and vice-versa. Sometimes people fear the consequences of knowing the truth – the so-called "ostrich effect" (cf. section 5.4.2), therefore they prefer not to investigate but this does not mean that they would not succeed. Quite the opposite.

According to Frankfurt, the case of bullshit is different, as the bullshitter is supposed to lack any concern or commitment to the truth-value of what he says. What turns out to be extremely puzzling is not the content, but his attitude. For instance, a liar voluntarily gets something wrong. But in doing so he conveys a certain commitment to the truth-value of what he claims. A bullshitter does not care about it. As just

---

[2] For a logical and epistemological treatment of bullshit, see for instance Carnielli [2010].

[3] On this, see also Carson [2010]. Carson claims that it is actually possible that one tells a lie while producing bullshit.

mentioned, a liar has a deceptive intention that can be detected. Whereas the case of the bullshitter is different. When a person believes *P*, she intends to believe *P* and this intention becomes meaningful to other people. In the case of the bullshitter, he believes without real intention to believe what he believes. So, what really defines a bullshitter is his attitude towards truth, or that he fails to be committed to truth. He simply does not care whether what he says is true or false, accurate or inaccurate.

This analysis on bullshit allows us to argue that bullshitters are basically not docile: they are simply careless about the beliefs they hold, an attitude which does not permit their knowledge to be passed on. Roughly speaking, what comes out of the bullshitter's mouth is *hot air* or *vapor*, meaning that the informative content transmitted is *nil*.

As Frankfurt brilliantly argued, a bullshitter *fakes* things, and his faking has important consequences. For instance, it may completely mislead their partner in a conversation. This is connected with the role played by second hand knowledge and its connection with docility. As argued by Simon, we do lean on what other people say. In Chapter 1, I pointed out that gossiping fallacies are cognitively successful, as they are based on making use of others as an information source. Trust, for instance, is not informatively empty (cf. section 1.2.1 devoted to presenting the *argumentum ad verecundiam*). One decides to trust another person, because she has reasons to do so. As already noted, there are a number of clues we make use of in order to consider a particular source of information (a person, for instance) as trustworthy or not. What happens then with a bullshitter?

A bullshitter does not really mean what she says she believes. She does not have any concern about the source of what she chooses to believe in. She just happens to believe. Thus information transmission becomes highly *difficult*. Here I come up with another fundamental difference to lying. As already maintained, a lie is not informatively empty, because people have various mechanisms for detecting it [Vrij, 2008, Chapters III and IV]. Our lie detector is based on our mind reading ability. Basically, we can guess that a person might be lying, because we know that we can all lie and thus we are able to read other people's intentions. Could we say the same about bullshitters? Do we analogously have a sort of *bullshit detector*? However trivial this question might be, our answer is that we do not.

Following Frankfurt, I claim that a bullshitter is defined by the kind of attitude he takes regarding the truth or that he exhibits no commitment regarding what he came to believe in. My take is that we can infer that he is bullshitting only because we are already familiar with (or expert on) what the bullshitter is talking about. The cues that are useful to us are related to knowledge of what he is talking about but, in the case of second-hand knowledge, these are precisely what is missing. This would be a kind of vicious circle, as we would need what we lack (namely, knowledge) in order to detect a bullshitter and bullshitting.

The inability to detect bullshitting is linked to another messy aspect of human beings: the epistemic bubble. As illustrated in sections 1.1.1 and 2.4.1, an epistemic bubble is a cognitive state in which the difference between *knowing that P* and *believing to know that P* becomes phenomenologically unapparent. The radical thesis held by its proponent is that this epistemic embubblement cannot be avoided by beings like-us. I may refer to bullshitting as a special case of epistemic embubble-ment. By definition we are embubbled with respect to a certain piece of knowledge. Embubblement allows us to take a sort of leap of faith, in that we should not believe we know that P, but actually we do. In the case of bullshitting the object of our em-bubblement is believing, meaning that bullshitters simply believe they believe, when they do not. Or as Frankfurt put it "[the bullshitter] misrepresents what he is up to". However weird this formulation might seem to be, it captures the very essence of bullshitting, which is not to care about the mechanism responsible for checking the plausibility of what one believes in.

Certain similarities to confabulating are worth mentioning to further explore the cognitive dimension of bullshit. In four words confabulations are false reports about memories [Hirstein, 2009]. Quite recently, a number of studies have been conducted in order to shed light on the very nature of confabulation. One of the most interesting conclusions that confabulation is due to a *reality monitoring deficit* [Schnider, 2001; Fotopoulou *et al.*, 2007]. This deficit is explained at a neurological level by the ef-fects of focal lesions on the medial orbitofrontal cortex [Szatkowska *et al.*, 2007]. Basically, confabulating patients lack the mechanisms enabling them to inhibit information that is irrelevant or out of date. For instance, they are not able to distin-guish between previous and currently relevant stimuli [Schnider, 2001]. As a con-sequence of this deficit, they are simply unable to control and assess the *plausibility* of their beliefs.

Some confabulations are extremely puzzling for their weirdness and they have already been recognized as symptoms of particular syndromes.[4] For instance, a pa-tient affected by Anton's syndrome denies being blind when they are, while those who suffering from Capgras' syndrome think that their relatives have been replaced by impostors. Amazingly, patients diagnosed with Cotard's syndrome think that they are dead.

More generally, confabulation emerges as human beings have a natural tendency for "coherencing" and filling in gaps. Confabulating patients have to face up to memories or perceptions that are basically false, but that they have accepted as true because of the reality monitoring deficit that they are affected by. The need for co-herence then creates explanations that result completely implausible and unaccept-able. In this sense, as in the case of bullshitting, confabulating is not lying but they are both affected by what has been called *pathological certainty* [Hirstein, 2005]. Basically, confabulators do not doubt, when they should doubt.

In bullshitting all these effects I have briefly surveyed are present, although they do not result from a deficit; they are brought about by the mindless attitude bull-shitters have about truth-value. Drawing upon Hirstein's phenomenological defini-tion of confabulation [Hirstein, 2009] I define bullshitting as follows. A person is bullshitting when:

1. She believes that P.
2. Her belief that P is ill-grounded.

---

[4] An exhausting list can be found in Hirstein [2005].

3. She does not know that P is ill-grounded.
4. She should know that P is ill-grounded.

To capture the difference between confabulation and bullshitting, points 3 and 4 are very important. In confabulating patients a person holds an ill grounded belief because of her neurological deficit. As already mentioned, she has a reality monitoring deficit that impairs those mechanisms inhibiting irrelevant or out of date information. If she were a normal person, she should know that P is ill-grounded. The case of the bullshitter is quite different. What overlaps between the two cases is that both of them should know that P is ill-grounded, when they do not. However, what does not overlap is the reason behind that. Bullshitters have no lesions preventing them from meeting any "epistemic" standard of truth. P is ill-grounded because of their careless attitude. So, in the case of confabulators they simply get things wrong, because they cannot discern relevant and up to date information from that which is not. Whereas bullshitters do not get things wrong, but, as Frankfurt put it, *they are not even trying*.

Once we distinguish between these two states of mind, then we analogically claim that the bullshitter is confabulating, as she holds a belief that she would withdraw or, at least, re-consider, if she tried to get things right.

5.4.2 The Ostrich Effect: The Limits of Docility

So far I have been dealing with docility as something virtuous and undocility as something vicious. Indeed, I have implicitly assumed a normative stance on docility, taking docile behavior as somehow "moral", and undocile behaviour as "immoral". I do not want to deny that this assumption has been made. However, while depicting the undocile individual, it is also worth considering the limitations of docility. That is to say, even if docile behavior is *somehow* good, on certain occasions it is not desirable or beneficial.

Docility underlies a number of activities concerning the sharing of information and cognitive resources. As just mentioned, it assumes a kind of ethical attitude that can involve avoiding information asymmetry, granting access to cognitive resources to as many people as possible, being committed to the general principle of clarity, and so on. All these prescriptions, however, do not take into account the simple fact that sometimes it is better to leave some things *unsaid*. Sometimes it is better to be undocile, that is, it is better not to disclose and then share information. As I will discuss in the following, the act of keeping silent – even with good reason – is indeed undocile.

I refer here to the so-called "ostrich effect" [Karlsson *et al.*, 2009; Brown and Kagel, 2009]. Basically, the ostrich effect is the tendency to ignore unpleasant information by means of avoidance and/or denial. The result is to remain silent over certain matters, thus blocking the possibility that a certain piece of information is made publicly available or acknowledged. This is also described – still in metaphorical terms – by the English expression "the elephant in the room", to refer to an object that everybody is indeed aware of yet no one wants to publicly acknowledge [Zerubavel, 2006].

There are a number of reasons why people might prefer not to publicly acknowledge that they have some problem. Consider, the case of some holocaust survivors who refused to share their experiences passing on the same attitude to their children and grandchildren. What they experienced in the concentration camps was so violent and horrific that they decided not to talk about it even to their closest relatives. And even when asked to explicitly mention some experience from that time, they continue to resist by resorting to the use of euphemisms, like for example, in the case of the holocaust, "unmentionable years" or "the war". Paradoxically, the recourse to euphemisms (that basically change the names of certain events) allows victims to refer to brutal experiences without actually mentioning them. Holocaust survivors are "silent witnesses", who might prefer not to share their experiences because it would be extremely painful and traumatic to disclose them.

One may opt not to talk about certain events not because of a trauma, but because it might involve fear or lack of confidence. For instance, in a family members may decide to keep silent about a member's drinking problem because they are afraid of the consequences or because they fear they cannot address the problem. Active avoidance is like a surviving strategy, when, for instance, people lack the resources to face a problem.

On other occasions, denial is just a matter of *tact*. We purposefully ignore a certain detail of our friend just because we feel we should. This may regard more or less trivial things like, bad breath, weight gain or hair loss. Disclosing and communicating certain information may irritate people or make feel them embarrassed. We save other people from *losing face*. This might be also be related to privacy and its ethical underpinnings. For instance, we may omit to say that our colleague's husband cheated on her causing depression, when the boss is going to assign a new and important task, because that could influence his decision. Here it is implicit that the communication of certain information is not ethically neutral, as it can promote malicious gossiping or, even worse, discrimination or mobbing. More generally, even in the absence of legal duty to privacy, silence can be a protective measure related to respecting people's ability to develop and realize their goals [Magnani, 2007c].

Denial may also acquire a social dimension, especially when it regards something that might turn out to be threatening for a group. For instance, some documents are *classified* by the government, because they contain information pointing to some vulnerabilities that might be harmfully exploited, for example, in a terrorist attack against inert citizens, or information with strategic value, for instance, on the stock market. In this case, investor interests are protected.

Silence or active avoidance may also regard the communication within a particular institution like the army. Recently, congresswoman Loretta Sanchez has called attention to the stories of female victims of sexual assault perpetrated while serving in the US army. Thanks to her initiative, a Sexual Assault Database has been set up and developed in order to encourage women to break the silence. Interestingly, a new "restrictive reporting option" has been introduced that allows women who have been the victims of sexual assault to get help (i.e., counseling and other treatment) without their command having to be notified.[5] In this case silence and active avoidance is explicitly supported by the law to protect the offended. This is an interesting example, since it shows how silence can be usefully deployed to break silence.

The ostrich effect can be seen as a particular case of self-deception, which stresses the relationship between language and self-deception. Indeed, silence is not inattentive, it is not an attitude which leads us to ignore something. It is a sort of *negative selection*, meaning that we simply prevent something from being transmitted to the public arena. As already argued, this is the undocile side of silence and denial. It can further be noted that in active avoidance, we simply refuse to extend our cognition and to consider some potential chances that are *over there* – distributed in the various external artifacts available in the cognitive niche. In the case of silence, we simply refuse to use language as *the ultimate artifact* to enhance understanding or to share potentially relevant information [Clark, 2006; Magnani, 2009]. As a matter of fact, language reconfigures a variety of cognitive tasks as it is a medium for "non-domain specific thinking" [Magnani, 2009, p. 163]. It also stabilizes certain experiences enabling important meta-cognitive abilities like having *thought of thought* [Clark, 2006]. Conversely, avoiding the use of language enables what I call *obliteration* (cf. section 2.6.2). Let me make an example in order to clarify this point.

As argued by Hirstein [2005], self-deception always involves tension or nagging doubts. Self-deception is the *result* of a process in which two inconsistent cognitions come to mind. For instance, consider a man whose wife has cheated on him. In this case, the self-deceptive process is activated or triggered because of evidence making him think that his wife is having an affair. He certainly experiences tension between the unhappy evidence he discovers and his belief that his wife *is* "faithful". The *initial awareness* of such a conflict is fundamental for the process of self-deception to be instantiated. It is only by means of manipulation that he is able to resolve the tension he is experiencing. Such a manipulation is not performed "internally", but it occurs in a "distributed" framework and in a hybrid way, that is, the initial awareness of a tension between the two conflicting cognitions is appropriately manipulated so as to favor obliteration. In the case of silence, it is the active avoidance of using language that helps to resolve the internal conflict. Initial awareness is literally switched off like a lamp promoting the process of embubblement (see section 2.4.1). This process is clearly undocile, because it is based on covering up potentially public information.

This undocile behavior more or less explicitly acknowledges the violent dimension of language. Silence and denial are explicit responses to this. Language – which is the product of massive cognitive delegation made possible by docility – turns into a weapon able to harm people. As argued by Magnani, language is rooted in a kind of *military intelligence* as a morality (and therefore potentially *violence*) carrier [Magnani, 2009; Magnani, 2011]. Basically, language transmits vital pieces of information that, once externalized and made available *over there*, acquire a kind of *semiotic agency* affording certain moral behaviors variously related to *expanding*

---

[5] The whole story is reported here: http://news.bbc.co.uk/2/hi/americas/8511010.stm

and, at the same time, *constraining* our action possibilities. Due to this creative role, language may promote or even create new inter- and intra-group conflicts.

I have another interesting example of how docility certainly has limits. Recently, the New York attorney general, Andrew Cuomo, started an investigation of the role played by eight banks in the attempt to fool rating agencies during the recent financial crisis of 2008.[6] More precisely, the investigation sought to understand whether the eight banks provided misleading information to rating agencies so as to make them overstate the value of mortgage securities, which ended up causing the collapse of the bank system worldwide in 2008. Mr Cuomo challenged the idea that overstating the value of mortgage securities was due to the agencies' incompetence, in fact, he claimed that the rating agencies were duped by the banks which tricked them by furnishing a misrepresentation of the current state of their securities in order to get more positive evaluations.

Interestingly, one of the techniques used by the banks to dupe the rating agencies was to *reverse-engineer* the computer models the agencies used to devise their ratings. Reverse-engineering was facilitated by the fact that the rating agencies made their computer models public. Thus the banks could trick them starting from answers and then working backward to get the intended result. In addition, the banks hired some of the analysts that devised the models.

This is a perfect example describing how excessive docility could bring about unexpected negative consequences. In fact, the rating agencies explicitly declared that they made their models public to increase the transparency of the rating process. On the one hand, making the models available was indeed a product of docility, as it contributed to rendering the process more transparent, given the various conflicts of interest rating agencies were accused of. On the other hand, the decision to make models public gave the banks a powerful tool to trick investors and the markets.

I have already pointed out how docility may promote misuses of language every time a private dimension to protect (for example, one's inner feelings, emotions, and memories) is publicly disclosed. That is, the effect of docile behavior is to make that private dimension sharable in an external structure over there so that other people can benefit from it as a source of additional chances for action and thinking. In the case of the rating agencies, the disclosure of computer models literally made an entire powerful niche publicly available that offered banks the chance to reverse-engineer the models and thus empower themselves to sell bonds.

To sum up, undocility also underlies those situations in which a person simply refrains from using an external object as a product of cognitive delegation basically for protection. As already mentioned, this acknowledges that cognitive delegations do come at a price. Furthermore, the unskilled or irreflexive use of external artifacts like language, for instance, can cause negative or unhappy outcomes that could have been prevented or avoided. In this sense, undocility may be an option to re-appropriate and re-gain control over the cognitive meanings that people have lavished on external things and objects.

---

[6] The whole story is reported here:
http://www.nytimes.com/2010/05/13/business/13street.html?pagewanted=1&hp

5.5 The Open Source Model as a Case in Point

Now, let me go back to docility. In this last section, I shall present the Open Source Model as a case in point describing *in vivo* docility.

Historically, what it is commonly labeled as Open Source (OS) embraces an astonishing variety of methods that cannot be reduced to one single approach. The great number of licenses under which work can be released is just one example demonstrating this fact. Free and Open Software, GNU and GPL licenses, Creative Commons, Copyleft, Open Standards, are just some of the different projects that belong to the Open Source galaxy. Here, I do not aim at examining all the differences between these approaches and ideas. Instead, I simply refer to the term "Open Source (OS) Model" as a general mode of knowledge transmission and creation that is based on one very simple idea: the source code of a software must be visible and editable so that it can be used, redistributed, changed, and upgraded by everybody.

What strikes us most about the various OS projects is the tremendous success they have gained during the last two decades. Linux is indeed the most well known case, but there is a number of other OS projects from web servers (such as APACHE) and office suites (such as OpenOffice) to script languages (such as PHP), from databases (such as MYSQL and POSTRESQL) to protocols (such as TCP/IP), that have now become leading products within their sector. Although Bill Gates forecasted that the hobbyists of the computer (as hackers were called by Gates) would have soon disappeared, the success of OS software and products is now widely recognized as a major event in the history of computing and business

Now, the point I want to develop here is whether "being open source" might derive from the way individuals process information, and relate to each other. What is the cognitive basis of individuals operating "open source"? Is there any social cognitive motivation for "being open source"?

5.5.1 A Matter of Cognitive Reliability

During the last few years economists and sociologists have provided stimulating accounts that try to explain the success of the Open Source movement (hereafter OS) and its rationale. Some authors [Raymond, 2001; Moody, 2001; Wark, 2004; Weber, 2004; Williams, 2002], for instance, pointed out that the radical innovation of the OS model was social rather than merely technical. To explain this idea, he introduced an illuminating metaphor that clearly depicts the culture and the values of the OS model. Most of the companies involved in programming he – argued – resemble what he called "a reverent cathedral building" with a rigid hierarchy. In contrast, the OS model is more like a "babbling bazaar of different agendas and approaches". No rigid hierarchies, no bosses, but very committed users that report bugs, and are also able to fix them and suggest alternative solutions or new problems to solve. On another note, Himanen [2001] provided a sociological account in which he compared so-called hacker ethics with Protestant work ethics and drew some interesting conclusions about the impact that this new radical approach may have on existing theories of business.

Although that sounds most appealing, these kinds of accounts do not hold water, because they fail to put forward any explanation about why *being open source* can also be extremely successful, from a cognitive perspective. For the main task is to investigate the cognitive reliability of the OS model and open up its cognitive kernel. Generally speaking, the main idea is that being open source may be something more than a business philosophy or a type of work ethic: it may also match a general trait of human cognition in the way it works and evolves. The point I want to make is the need for a cognitive account of the success of the OS model. Generally speaking, the OS movement deals with information and knowledge transmission: therefore the way it manages, organizes, and extends cognitive abilities to cope with programming becomes a crucial aspect that cannot be neglected.

In the following section, I describe the four dimensions in which the notion of docility may be a valuable candidate in explaining the cognitive relevance of the OS model.

5.5.2   The Docile Hacker

The very idea of the OS model is that the source code of a software must be visible and editable so that it can be used, redistributed, changed, and upgraded by everybody. Now, the point I want to make in this subsection is that sharing code is a product of docility.

Source code is not just a block of bits that saves time for those who, fortunately, can use it. Source code is a cognitive repository that stores ideas, problems, trials as well as errors, solutions, and it may suggest alternative views. If that is correct, then sharing code contributes to releasing a large body of knowledge and information that drastically modifies how other people (in this case, hackers) can learn, solve problems, and more generally accomplish a cognitive task such as that of making up computer programs. As Raymond put it "you often don't really understand the problem until after the first time you implement a solution" [Raymond, 2001, p. 25]. In doing this, hackers lean on various external resources (in this case, the source code written by others) that become a major basis for their cognitive work and performances. That is exactly what docility is all about. That is, in writing and then sharing the code, hackers are continuously involved in a "smart interplay" between their brain and the environment that is facilitated and enhanced by a tendency toward external resources: that is docility.

Indeed, in the case of proprietary software, programmers share code and, to some extent, they are docile as well, because they take advantage of others' improvements. However, docility is limited by the narrow boundaries of the company they work for: nobody else can access the code. In contrast, hackers can potentially rely on thousands of people all committed to the same problem.[7] An example of this enormous

---

[7] In a famous paper of his, Bill Gates (1976) argued that hobbyists could not have built up reliable and stable computer programs. They must have been paid for doing such a good job. On this note, he wrote: "Without good software and an owner who understands programming, a hobby computer is wasted. Will quality software be written for the hobby market?".

potentiality is given by the high reliability that open software guarantees to the user. As Raymond wrote "many eyeballs tame complexity" [Raymond, 2004]. As a matter of fact, Microsoft products (from computer servers to PCs) are much less reliable than Linux in terms of security, scalability, performance, compatibility, stability, and so on. In proprietary software companies, the fact that docility is limited jeopardizes all the cognitive benefits provided by docile behaviors. For instance, peer review, that is indeed one of the most successful factors leading to software reliability, is dramatically reduced. In fact, the peer review principle is based on the possibility for everybody to check each other's work without limitations of any sort. None can hide his/her work and prevent others from criticizing it. In this sense, secrecy is the enemy of quality and it can be regarded as highly anti-docile behavior.

I argue that code-sharing contributes to releasing a great portion of knowledge that drastically shapes the cognitive task hackers face. The same can be said for another feature of the OS model, not in this case connected to inanimate resources (the code) but animate ones, that is, other human beings. In this case, docility is crucial to making use of those cognitive resources embedded in social channels. That is, hackers are docile in the sense that they do not simply work on the same piece of code: they build up communities of practice in which learning from others and then teaching what is experienced becomes a major trait in the way knowledge is transferred and developed. In this sense, the social dimension turns out to be a significant cognitive source for their work. Peer review is indeed an example of this kind, as briefly discussed above: people that get involved in an open project release their work openly to other hackers that in turn provide them with suggestions or improvements or simply test their distribution. Cooperation is therefore a direct consequence of the way they work, not only an *ethical option* [Himanen, 2001].

Docility is also displayed in the ubiquitous use of social tools such as forums, chat rooms, mailing lists, newsgroups, newsletters, etc. As a matter of fact, for any open source project there is a community of practice and learning. Hackers and developers are allowed to exchange information, solutions, suggestions, know how, etc. As a matter of fact, most of the activities concerning software development are managed and organized through the Internet. Usually, open source projects start out from a person or a group of people that stumble over a series of unsolved problems. Then, they post some information about their problems on a Website or a mailing list and try to get some help from other hackers. This gives rise to a community in which hackers can freely cooperate on the project or simply get an idea of what is going on. Thus it is not surprising that historically the success of the OS Movement was largely due to the creation and implementation of tools that enabled distance communication.

In this sense, forums, chat rooms, and the like are cognitive mediators that encode and then release a great portion of resources embedded in social channels and facilitate knowledge transmission at the same time.

Source code can be considered a cognitive repository that is open to everyone who wants to modify or simply re-use it. But that is not the whole deal: the success of the OS Model is also related to the tremendous developments that it brought about. For instance, when the first version of Linux OS came out in 1991, it consisted of only 10.000 lines of code. Just after 7 years, it was made up of more than one and half million lines. What does this mean? It means that the OS Model is not only about sharing code, but it is also a development model in which progress is really made possible by the thousands of hackers involved in various open source projects. That is, hackers do not only re-use and share code, but they are committed to sharing any development or contribution that may improve the quality of a software. According to the GNU General Public License, any modification made upon every single piece of code must be released, since everybody must give the recipients all the rights that are given to him/her. Here again the role of docility is crucial in describing the cognitive relevance of this attitude; hackers are docile in the sense that they opt to publish their improvements for further inspections, to fix some bugs or add new features.

Standard-fidelity is one of the most interesting aspects concerning docility, as already pointed out. The cognitive relevance of following standards or/and standardizing one's own work is as follows: first of all, using standards makes information and knowledge transmission much easier. This is true for humans and also for machines. Consider the case of formal languages comparing with informal ones. Mathematicians, logicians, etc, try to make up ways of communicating and transferring knowledge that are transparent (standardizing) as much as possible to overcome the ambivalence and ambiguity of natural languages such as English, French or Italian. Secondly, having standards also makes it much simpler to compare different claims. Consider a scientific experiment: here scientists follow certain standardized procedures that clearly display results and the way to test their presumed validity. Very often the incommensurability between theories is due to the failure to apply set standards when measuring the different claims and to then decide upon the best method. Thirdly, standards facilitate further developments. Here again, standardized procedures lead to results that can be understood more easily and shared better by the community of practitioners (scientists, mathematicians, and so on). Now, let me turn back to hackers and the relevance of standard-fidelity for the OS model.

Usually Open Source is viewed as something related to software development, whereas standards regard common agreements that allow communications between different means [Krechmer, 2005] From an analytical perspective I do not find reasons to reject this distinction.[8] But from the hacker's point of view things starts blurring. The main motivation that stands behind the very idea of the Open Source is to keep the source code open and available to everybody for inspection and modification. Therefore, it is ultimately committed to enabling people to use and exploit all the cognitive functionalities that a software can give, without any restriction. If that is correct, then building up standards that amplify interoperability (interaction), cross-platform compatibility, usability, and so on, is a part of the OS kernel.

Now, focusing more on standard-fidelity in computing, I find two main levels at which it operates. The first one regards the kind of standard-fidelity displayed by mathematicians. As argued above, mathematicians play their game by the rules of the discipline that are not personal or subjective. The same happens in making

---

[8] For an interesting debate about the distinction between Open Source and Open Standard, see Chawner [2005].

software. Since in the Open Source galaxy a piece of code should be easily shared and modified by all, some basic requirements must be met to increase re-usability. These basic requirements regard, for instance, writing code (consistency and clearness, for instance). Some of them are also related to releasing pieces of code under some open source license. For instance, GNU Free Documentation License regulates verbatim copying, modifications, the documentations to release with the code, and so on.

The second aspect of standard-fidelity explicitly concerns the increasing of open standards as a major opportunity to disseminate and distribute knowledge and cognitive capabilities. There are many projects concerning open standards. Among them, it is worth citing the case of Open Document Format (ODF) developed by the OASIS11 industry consortium; ODF has been recently approved by ISO12 (International Organization for Standard) as the first standard for editable office documents. Another well-known example is the World Wide Web Consortium, primarily devoted to developing standards for the Web.

5.6 Concluding Remarks

In this chapter I presented the notion of docility. In my view, docility is supposed to facilitate the delegation and exploitation of cognitive chances secured to cognitive niches. Docility is thought to be a fundamental behavioral correlate that makes the extension of human cognition possible, promoting the process of its de-biasing. In the second part of the chapter I also dealt with the problem of undocility, which is not simply to be considered a selfish behavior, but also an active strategy protecting human beings from the negative consequences various eco-cognitive externalizations can have on them. In the last section of the chapter I have shown how explicative the notion of docility can be with relation to the case of the Open Source Model. In the next and last chapter, I will change setting offering the reader an alternative approach to morality and ethics based on the distributed approach.

Chapter 6
Seeking Chances: The Moral Side

Introduction

The present chapter could be considered an appendix devoted to application of the approach so far developed to morality and moral reasoning in general. I claim that the mechanism underlying chance-seeking activities may encompass some important features of moral reasoning. This chapter is not going to put forward any moral theory; rather, it aims at providing a cognitive framework for morality.

In section 6.1 I will argue that from a behavioral point of view morality is about empathizing with other people. This structural aspect of morality is developed taking advantage of the notion of *moral proximity*. Basically, moral proximity will be considered one of the most fundamental variables prompting an empathetic response to others. That is, the more a person is proximal to a certain event, the more he will consider himself morally committed towards the person or people involved in it.

Coherently to the framework developed in the previous chapters, in section 6.2 I will argue that moral proximity can be extended by means of external resources. More precisely, I will articulate the idea of *distributed morality* recently introduced by Magnani [2007c]. According to this hypothesis our capacity to comprehend a certain event as bearing moral concern can be enhanced by leaning on various external objects, which are apparently inert from a moral point of view, but that – upon occasion – can help us uncover additional information which directs our moral decision-making process.

In the last section I will present the notion of the *moral mediator*, in a discussion of the philosophical issues surrounding the agency of technological artifacts. In doing so, I shall consider the Internet as a case in point to provide a better picture of the idea of distributed morality.

E. Bardone: Seeking Chances, COSMOS 13, pp. 125–143, 2011.
springerlink.com   © Springer-Verlag Berlin Heidelberg 2011

6.1 Moral Proximity as a Leading Factor for Moral Understanding

6.1.1 What Is Moral Proximity?

In his seminal work on moral philosophy, the father of modern economics Adam Smith [2004] argued that morality is connected to *sympathy*. In his own words:

> As we have no immediate experience of what other men feel, we can form no idea of the manner in which they are affected, but by conceiving what we ourselves should feel in the like situation [...] our senses will never inform us of what he suffers. [...] it is by the imagination only that we can form any conception of what are his sensations. It is the impressions of our own senses only, not those of his, which our imaginations copy.

We cannot have direct experience of what others feel, but by means of imagination we can figure out how a person feels and then have a moral response to him/her. Sympathy is the kind of *bounding mechanism* through which we place ourselves in another's shoes. In ethical decision-making theorists regard sympathy as belonging to a crucial step called *moral recognition*.[1] Most of time people fail to morally respond to a certain event or situation, not only because they lack rules to apply, but also because they are not able to recognize, understand or even feel it as a moral issue, that is, an issue that requires a moral commitment [Watley and May, 2004].

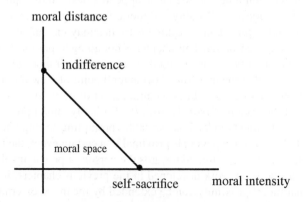

Jones [1991] argued that a moral issue can be represented in terms of its *moral intensity* (see the graph). As he put it, moral intensity is a "construct that captures the extent of issue-related moral imperative in a situation" [Jones, 1991, p. 372]. Building on Rest's previous work, Jones developed a model of moral intensity described by six main components that in turn can dynamically interact with each other:

---

[1] Rest [1986] introduced a four-components model for individual ethical decision-making according to which a moral agent must a) recognize the moral issue, b) make a moral judgment, c) resolve to place moral concerns ahead of other concerns, and d) act on the moral concerns. We do not necessarily take that for granted; I simply agree with him in recognizing a preliminary step called *moral recognition*.

1. magnitude of consequences;
2. social consensus;
3. probability of effect;
4. temporal immediacy;
5. proximity;
6. concentration of effect.

The model is very detailed and it is a valuable starting point from which to provide another alternative, but similar model. Jones's moral intensity components can be grouped into three main primitive elements:

1. *moral proximity* (or moral distance) that subsumes 4) and 5),
2. *consequences of an action* that subsumes 1), 3) and 6),
3. *social consensus*.

Let me make two points on that. First of all, we do not find the third element relevant, namely, social consensus. According to Jones, social consensus is defined as "the degree of social agreement that a proposed act is evil (or good)" [Jones, 1991, p. 375]. The point I want to make is slightly different; although people recognize the ethical valence of an abstract situation, depending upon the actors who are actually involved in it, the perception of a moral commitment in a given concrete situation can drastically vary from case to case. The different response does not depend on the normative content of that situation which, conversely, remains the same. In this sense, there may be social consensus on a certain issue (perceived or not), but no response, because I maintain that empathizing comes first.

The second point is connected to the relationship between consequences and moral proximity. From a genealogical point of view, I posit that moral proximity comes first, whereas consequences are considered only secondarily. That is, the magnitude of consequences, concentration of effects, and so on, come to our attention – and are therefore processed – only after the process of empathizing has taken its course. Having set the scene, I introduce our hypothesis, which connects moral proximity to presence or absence of moral response towards a certain event. More precisely, I claim that the more distant a person is from a certain event, the less he/she will consider himself/herself morally committed towards the person or people involved in it. Conversely, the more an event is perceived as proximal to a person, the more she/he is morally committed. It is worth noting that moral proximity is not a normative claim for or against a certain policy or decision, in this sense, I agree with Singer who explicitly stated that "the fact that a person is physically near to us does not show that we ought to help him rather than another who happens to be further away" [Singer, 1972].

Spohn [1996] added that the sense of belonging to a given moral community is relevant. That is, from a cognitive perspective I recognize the ethical valence of a given event, if we *recognize* those who are involved in it as belonging to our moral community and the boundaries of our moral community affect our moral response (cf. the issue of group selection is dealt with in section 3.5).

6.1.2   Some Evidence on the Relevance of Moral Proximity for Moral Engagement

In a famous series of experiments Milgram [1974] reported that obedience to an authority tended to diminish, once the subject was brought closer to the victim. Milgram argued that *empathic clues* such as voice feedback, touch proximity, played a crucial role in perceiving moral commitment towards a person or a group of persons.

Bandura [1999] has provided an interesting theory called *moral disengagement*. He argued that people indulge in inhumane conduct, not necessarily prohibited by the law, because they simply disengage themselves from the actions they carry out. Moral disengagement involves, for instance, the *displacement of responsibility* by minimizing any role as agent in the situation, the use of *euphemistic labeling* (for instance, a war can be sanitized, if it is called a preemptive war rather than an aggressive one, or a bomb becomes "intelligent"), and the dehumanization of victims [Osofsky *et al.*, 2005]: as Bandura put it, "[o]nce dehumanized, they are no longer viewed as persons with feelings, hopes and concerns but as subhuman objects" [Bandura, 1999, p. 199]. This last point supports our hypothesis: dehumanization is less likely, if some empathic contact is established.[2]

The idea that the more distant we are from a person, the less we will care about them is supported by neurological data [Moll *et al.*, 2002; Greene, 2003]. Greeene [2003] conducted a brain imaging study in which subjects were asked to respond to this dilemma. Suppose a person A is driving a car when he notices a man dying on the ground. He jumps out of the car and helps the man. Now, suppose that a person B receives an email in which he is invited to send 100 euros to support ten thousand starving children in Africa. Although these two cases are similar from a moral perspective, we would surely blame the first person, if he did not help the dying man, whereas we would not do the same, if the second person did not send the money. It is worth noting that these two cases have the same normative content. Technically speaking, they are *isomorphic*. The issue here is simply about helping a person or a group of people that are in danger, respectively, the man dying on the ground and the starving children in Africa.

Greene [2003] reported that situations similar to the first case statistically involve more brain activity in those areas related to emotion and moral cognition. He argued that from an evolutionary perspective helping people is is immediately beneficial, that is, it is conjectured that humans have developed a mechanism to make quick moral decisions that involve others close to them, because it contributes to increasing the fitness of all. This mechanism can also be regarded as *short-distance* altruism. Conversely, the second case is an example of *long-distance* altruism [Gazzaniga, 2005], because we are not facing the person who needs help, and the distance involved does not lead to the lighting up of any neurological (and/or instinctual) mechanism related to reciprocity or altruism. The kind of altruism which

---

[2] An alternative view on moral disengagement is provided by Magnani [2011]. Magnani argued that the process of moral disengagement is actually a moral *re-engagement* in another morality.

our brain is wired up to reflects the kind of environment our ancestors faced. More-over, it is worth noting that helping others in the wider environment – like that of our ancestors – would have implied considerable personal sacrifice [Greene, 2003]. In this sense, the bounding mechanism selected by evolution could have been designed for *hot* situations and not for those in which relatively modest sacrifices are required (like the example of sending 100 euros via email to the starving children in Africa).

6.1.3 Moral Proximity Can Be Extended and So Can Our Moral Understanding

As mentioned above, the bounding mechanisms that human beings posses internally are limited, a fact that is visible in the dilemma illustrated above. Following Greene's conjecture, our bounding mechanisms reflect the kind of environment inhabited by humans for thousands of years. However, it is worth noting that the evolution of cul-ture, and lately that of technology, have dramatically changed the familiar landscape of our moral sense. Following what was argued in Chapter 3, the activity of cognitive niche construction has clearly had an impact on our moral understanding. Paintings, language, and the new forms of communication invented in the last two centuries (from cinema to the Internet), have had tremendous effects on moral distance and the way humans can now perceive moral needs and commitments [Russ-Eft, 2004]. It is worth noting that a website or a picture are not part of our brain-mind sys-tem. That is, there is no neuronal mechanism that can encode information the way a website or a picture does. In this case our internal capacities are simply bounded. However, working in concert with external resources (a picture or a website) we may dramatically enhance our moral performances. The famous motto "a picture is worth a thousand words" can be analogically applied also to ethics and moral cog-nition. As mentioned at the beginning, Adam Smith considered imagination as the faculty deputed to moral recognition. But can imagination be extended or not? And what about its effectiveness?

A positive answer to these questions has been provided by Magnani [2007c] who introduced the idea that morality is distributed. The core of the distributed morality hypothesis is that our capacity of seeing and understanding moral values is partly distributed across the things (animate and inanimate) we cope with on a daily basis. That is, external resources (a metaphor, an image, a book, a toy, a website, a friend or a stranger, and so on) work as *moral mediators* since they can uncover informa-tion, alternative perspectives and hypotheses that otherwise would remain unknown. In doing so, they contribute to the pondering over, formulation, and/or direction of our moral response and judgment. In sum, moral mediators represent a kind of re-distribution of the moral effort through managing objects and information in such a way that we can overcome the poverty and the unsatisfactory character of the moral options immediately represented or found internally.

What kind of impact moral mediators have on moral distance/proximity is now under investigation. A second group of questions is related to how the perception of being morally distant from a certain event may vary upon occasion. My second hypothesis can be summed up in that way: the exploitation of external resources significantly changes the perception of the moral distance of an agent towards a moral issue.

I argue that moral intensity is always mediated by *moral mediators* as I will detail in the following sections. That is, recognizing an issue as a moral issue depends on the capacity to bring the people involved in it closer to us. This process of bringing oneself closer to the victim is mediated by some structure. I use this term following Hutchins' definition, that is, a mediating structure (or simply a mediator) "mediates the relationship between the performer and the task [. . .] I will view it as one of the many structural elements that are brought into coordination in the performance of the task" [Hutchins, 1995, p. 290].

6.2   The Morality of Everyday Things

Several studies on distributed cognition have pointed out that the environment is a mediating structure filled up with various cognitive resources that can be picked up and made use of, on occasion. External resources provide us with additional computational capabilities, because they allow us to perform those actions vital to us. External resources are also memory stores, because they can encode information content and release it when needed. Databases are examples of this kind, since they store and then retrieve a large amount of data that we could not manage alone.

According to that view, almost all human performances are hybrid in the sense that they are brought about by various interplays between individuals and external objects. Following this idea, in 1988 Donald Norman published a book under the title *Psychology of Everyday Things*, in which he argued that most of time people's faults lay in the design of the things that they daily use. That is, people are not always to blame for their mistakes, because things are often designed as if they were built to cause errors. Therefore, the psychology of everyday things is metaphorically the discipline that should shed light on the interaction between humans and external objects in order to enhance our pre-existing capabilities and/or lessen their possible negative impact.

The idea I shall try to develop is that the same can be argued about morality. More precisely, I shall argue that everyday technologies (i.e., computers, the Internet, etc.) drastically modify our capacity of coping with all those situations that involve some moral concern. Accordingly, I may speak not only about the psychology, but also the morality of everyday things.

It is worth noting that robots and computers that are capable of acting morally are not currently available. They are still the stuff of science fiction, like that written by Isaac Asimov. No computers have intentions or desires and, indeed, they cannot act morally in a way similar to humans. So why should we care about the morality of computers and other technologies?

Although intelligent or moral machines are mere intellectual fiction, the quality of the interaction between humans and technology has become increasingly complex with computers, for instance, dramatically increasing the moral impact of external supports on our lives. New moral issues and concerns have arisen. Consider, for instance, our privacy: computers have made it possible to gather and store such a large amount of personal data and information that the notion of ownership has been drastically modified [Magnani, 2007c, Ch. IV]. New technologies have also provided us with new tools to cope with pre-existing problems. Just think of the tremendous opportunities offered by the Internet for democracy and education. Both *e-democracy* and *e-learning* are currently changing the way people organize their political activities, transmit knowledge and teach.

New fields of study have emerged. Consider, for instance, *computer ethics* and, more recently, machine ethics. The first brought up a series of discussions and debates that traditional ethics and moral philosophy have completely discarded as marginal or too applied [Bynum and Rogerson, 2004; Moor and Bynum, 2002; Floridi, 1999; Johnson, 1994]. *Machine ethics* comes from the AI (Artificial Intelligence) tradition, and it aims to provide an ethical dimension to technological devices [Anderson *et al.*, 2005]. Many emerging topics are being elaborated and discussed, like for instance, the interaction between artificial and natural intelligence systems and machine-to-machine communication and cooperation. Machine ethics is also involved in building artificial systems that are able to assist humans in ethical decision-making.

Although these fields of study have furnished valuable and innovative contributions, they restrict their moral contributions to examination of the possible misuses of technology. They somehow assume that technology is just a means that is "external to the context in which it operates" [Buchholz and Rosenthal, 2002, p. 48]. More precisely, there is a lack of understanding about technology's cognitive and moral dimension. A number of questions have been discarded, even if they are crucial in dealing with technology. For instance, why do computers really matter? What is their cognitive role? How do they contribute to our moral understanding?

As mentioned in Chapter 3, recent studies on distributed cognition have pointed out that computers (and other external resources) are inherently a part of the human cognitive system, as we continuously modify the cognitive niches which we live in. Clark [2003] argued that we are *natural-born cyborgs*, because we constitutively exploit the cognitive resources embedded in various external objects and supports (computers included). In fact, that is the way human cognition works and evolves. I maintain this conceptual framework can lead us to a different perspective on the moral relevance of external objects like computers and other computational machines.

Recently, Magnani [2007c] introduced a new paradigm that profoundly alters the familiar landscape of ethics and its relationship with technology. Magnani argued that morality is a distributed phenomenon, like cognition, morality is distributed in the sense that the capacity of acting morally or seeing moral entities also depends upon external resources and the way we manipulate them.

6.2.1 The Idea of Distributed Morality: A Cognitive Framework for Ethics

From a cognitive perspective, I maintain that human beings are problem solvers [Simon, 1947; Simon, 1955]. People are continuously engaged in solving problems all day long, every day. Some of them are more trivial, such as choosing which clothes to wear or buying a car. Others are much more complicated: choosing which university to attend, changing job, deciding whether to marry Suzie or John, whether to invest in a Chinese corporation, to fund a charity or to support Greenpeace. Some of these are labeled as moral problems, since they involve other people: their health and happiness and everything that concerns their life as human beings.

Ethical deliberation, and morality in general, can be considered as a problem solving activity in which people try to apply pre-existing solutions and/or generate new ones to complete the various tasks they face. However, as in any kind of problem solving activity, ethical deliberation is based on intrinsically incomplete information, because it is impossible for anyone to be aware of every fact related to any given subject. That has important theoretical implications:

1. having incomplete information means that our deliberations and decisions are never the best possible answer, but they are at least satisficing;
2. our conclusions are always withdrawable. That is, once we get more information about a certain situation that involves some moral concern, we can always revise our previous decisions and think of alternative pathways that we could not "see" before;
3. a great part of our efforts in solving a moral task are devoted to elaborating conjectures or hypotheses in order to obtain more adequate information.

Within this framework, conjecturing is essentially an act that permits us to manipulate our problem, and the representation we have of it, so that we may eventually acquire more "valuable" data. In this sense, I maintain that morality is manipulative in its essence, because it deals with changing and manipulating the world in order to overcome the unsatisfactory character of the options that are immediately available.

In generating conjectures and hypotheses, I claim that the role of external resources (i.e. computers or other forms of technology) is fundamental. More precisely, the continuous interplay between individuals and their environment is one of the most distinctive traits of moral cognition and morality in general. Here the main thesis I put forward is that this interplay is a kind of *semiotic activity* in which our minds are continuously extended into the external world [Magnani, 2006a]. This process can be split into two:

- people externally reproduce something that they would usually only have within the isolated brain and thus make it more visible;
- once they have externalized their thoughts using external objects, people can work on them and develop new concepts and new ways of thinking. The entire cycle is called the "externalization process" (cf. 3.1.2).

During the externalization process individuals create something that exists without their brain. This process is called mimetic, because individuals use external supports to reproduce what occurs inside their private consciousness and, in turn, make their thoughts *easy-to-share*. One of the most common mimetic activities is writing; writing is a mimetic activity, because people represent and reproduce thoughts and ideas by another means (the sheet of paper). There are many other activities of this kind that involve computers. As mentioned above, software engineers and computer scientists have recently started caring about the moral dimension of machines and computers. For instance, designing a software agent capable of assessing the moral ramifications of courses of action has become an amazing new challenge. Implementing such an artificial system is, indeed, a mimetic activity, since engineers and scientists try to artificially reproduce behaviors and processes.

The example of writing is also interesting in another way. Once our thoughts have been secured to an external support (the sheet of paper), we are able to think and operate on them in a way that would not otherwise be possible. As a matter of fact, we cannot re-read our thoughts, because they are fleeting and immediately fade away. But, once written, we can use the sheet of paper as a creative external representation and perform some activities otherwise impossible [Magnani, 2006a]. More precisely, external supports allow individuals to re-project their own thoughts so that they can uncover hidden information and new concepts (cf. section 3.1.2). In this sense, external objects take part in creating and finding room for new ideas and perspectives.

Now the question is how can we take processes like this into account if we consider morality as related only to the application of rules, imperatives or guidelines? As the example shows, ethical deliberation and morality are expressed not only in words at a verbal/propositional level [Johnson, 1993] but also though model-based and "through doing" processes [Magnani, 2007c]. This is the basic point I want to stress here.

As mentioned above, people often exploit external supports (for example, language but also technological innovations) to enhance their moral efforts in a completely tacit fashion as the example demonstrates. I can distinguish two main types of moral behavior employed by human beings. The first type is related to selecting the most appropriate course of action from a library of pre-existing behavioral templates that can be considered as automatic responses. The second type regards all those situations in which humans do not rely on pre-existing solutions, but invent new ones. In the following section I will provide some examples of moral templates, and I will discuss the creation of new ones.

Generally speaking, a moral template is a set of actions and decisions that have been successfully tried and tested within a society or culture. In this sense, it represents a variety of solutions that become culturally and socially accepted and that on occasion are deployed and thought to be the best course of action in certain circumstances. Some of these solutions are linguistically encoded in guidelines and imperatives. The Ten Commandments are an extraordinary example of moral templates, because they provide moral options and solutions that may help people in many circumstances. Moral theories themselves are examples of this kind, because they provide a highly theoretical guide to various aspects of everyday life. Utilitarianism, Kantianism, and the social contract are thus ways of interpreting the world in a moral sense that give us explicit, coherent, and consistent reasons for our actions [Thagard, 2000].

Many other templates are not explicitly laid out or expressed in a sentential way, but remain embodied in actions and bound up with various external structures and configurations. The case of the email I presented above is an example of this kind: writing is indeed an experienced and powerful template that can be used to reflect upon some complicated issue and/or to manage our emotions and feelings in order to assess whether they are appropriate or not [Harris, 2004; Love, 2004; Wheeler, 2004].

Another example of a moral template that is not sentential, but embodied in an external structure is represented by the various institutions that we find in many societies and cultures. The institution of the family is an example of this kind. The family can be considered as a template that groups various successful solutions to some problems related to survival, but also to parental care, the role of elderly people, property, the division of labor, and many other issues that can be the source of problems and conflict among human beings.

The nature of these templates is highly conjectural. As a part of problem-solving activities, their validity rests on the fact that they are successfully experienced and transferred to others as cultural inheritance. But they still remain retractable and open to improvement.

As noted above, sometimes pre-existing templates are not adequate to solve the problems that we face. As a matter of fact, templates themselves were once invented because the options available at a certain moment were not adequate. First of all, moral innovations sometimes represent a radical revolution compared to past templates. Let me consider the case of democracy. Democracy represented an amazing moral innovation, compared with pre-existing forms of government. It brought into existence a series of moral entities that were totally neglected before its advent. For instance, the notion of citizenship gives a moral and equal status to everyone, a classic example being: *liberté, egalité, fraternité!* In other terms, the radical moral (rather than social and political) innovation was that every citizen had been appointed with certain basic rights that the State could not take away. In this sense, the idea of democracy creates morally intelligible entities, e.g. citizens. The moral status of human beings dramatically changed after the modern democratic State had been created, and things changed again when women were first allowed to vote. These are examples of how morality changes in history, and, with it, the objects that are given moral meaning.

Secondly, moral innovations may arise from pre-existing templates that are occasionally revised and modified. However, the process of revising old moral habits and concepts may be extremely problematic. Consider, for instance, the case of gay marriage. Its proponents attempt to solve some conflicts related to extending a set of rights also to gay couples by modifying the entire institution of the family. This issue can be highly controversial, because its opponents argue that the traditional family is a template that has been corroborated to solve certain problems, but not others. A precautionary principle is thus advocated. In some countries, the issue of gay marriage has been solved in a different way, gay couples being able to draw up a contract regarding their relationship in the same way that any other unmarried couple can. In this case, a new and different moral option is created by modifying, but not replacing, a pre-existing template.

It is worth noting that not all moral inventions become widespread prototypes or solutions. Some of them can be occasionally employed by a lone person or an isolated group, but soon discarded. That can be true, whether they were successful or not. First of all, because they can be immediately replaced by better ones. Secondly, because they cannot always be reproduced and/or transmitted. A new moral idea might be connected to a specific situation and context that cannot be replicated somewhere else. There have been plenty of moral and social experiments that aimed at reconfiguring the entire Western way of life but some of them failed as general revolutionary movements, because they were strictly linked to a specific historical moment. Once social and political circumstances changed, their moral appeal soon disappeared.

Finally, the failure or the success of new moral ideas also depends upon society and human decisions. Communities can adopt and discard ideas and innovations for various reasons that can independently be social, political or economical.

All these examples I provided point to the conclusion that morality is a manipulative and "through doing" activity in its essence. It aims at manipulating and reconfiguring pre-existing ideas to solve some problem related dealing with other human beings. As shown above, a great amount of our efforts are devoted to building various moral behavior templates that help us solve some specific problem. In doing this, the exploitation of external resources is crucial. Morality is fostered and enhanced by continuous moral delegations in which we transfer a large amount of ethical knowledge to various external and mediating structures, such as language, theories, institutions, and technological artifacts. In turn, what we have delegated to external structures (e.g. democracy, or democratic institutions, political representatives, pools, statistical services, etc.) could help us to generate new ideas and a re-projecting activity is thus carried out [Magnani, 2006a]. It is a "re-projecting" activity because we introduce information each consecutive time, as it has been modified outside our brain (cf. section 3.1.2). In this sense, individuals create new knowledge through the exploitation of external supports/resources. I maintain that morality is a distributed phenomenon [Magnani, 2006b]. That is, we cannot refer to morality as something that happens only within the human mind, but it is somehow distributed over a set of external resources and internal capabilities.

The framework I briefly detailed in this section constitutes a starting point to better understand the moral role and relevance of technological artifacts such as computers.

6.2.2 Epistemic and Pragmatic Actions: The Moral Side

In the last section, I pointed out that morality is a distributed phenomenon, since people rely on external resources to make decisions or generate new ideas. In order to avoid possible misunderstanding about this claim, it is worth citing the distinction between pragmatic and epistemic actions [Kirsh and Maglio, 1994]. Generally speaking, a problem can be defined by an initial state, a goal state and a set of operators (or mediators) that allow transformation of the initial state into the goal state by a series of intermediate steps.

These intermediate steps, that I will call hereafter actions, can be grouped into two main categories: pragmatic and epistemic. By the term "pragmatic actions" I refer to all those intermediate steps that alter the world to achieve some physical goal or other physical intermediate stages. For example, if one wants to be refunded for a certain purchase, he might have to fax the receipt. The action of faxing the document is a pragmatic action because it brings one closer to the goal state, namely, being refunded. In contrast, epistemic actions are all those actions that alter the representation of the task one is facing. A child that shakes and feels their unwrapped birthday present to guess what is inside is a fair example of this kind; the action of shaking unearths additional information that makes guessing less blind. In this case, the world is not strictly changed but what is changed is the representation we have of the problem. Accordingly, epistemic actions can also be regarded as task-transforming representations [Hutchins, 1995].

Analogously, the same can be argued for moral situations. Indeed, a moral task can be considered a problem-solving activity, as I discussed above. If this consideration is correct, the impact of technology on ethics and morality is twofold. First of all, technology deals with all those actions that can help us to pragmatically enhance or diminish our moral effectiveness. Secondly, computers and machines also constitute external representations that transform the moral problem we face and thus help to solve it.

Let me start with pragmatic actions. Consider, for instance, the case of Google. Google, as one of the most powerful web search engines, permits us to go through a huge number of web pages. Recently, some ethical concerns related for instance to privacy have arisen. Some years ago Google^{TM}released a new webmail service called Gmail (http://gmail.google.com). According to its inventors, the appeal of Gmail^{TM}is that it comes with built-in Google technology, and therefore it allows users to easily search their mailbox for messages . In addition, Gmail provides users with the possibility of receiving advertisements and information that are relevant to them or their messages. Neither pop-ups nor untargeted banner ads are sent. Although this is advertised as a special feature that no other email services offer, some problems related to privacy immediately arise.

Now, consider the following example. Suppose a person A writes an email about apple pie to a person B. Since Gmail provides users with personalized ads, B can receive information for apple pie recipes alongside his email [Batelle, 2005]. This can happen for whatever one writes, which might encompass political or sexual orientations, hobbies, news, and so forth. As suggested by Batelle, it is "as if someone at Google was really reading ... email, then choosing the ads that should accompany it". We know that is not true. Nobody at Google is actually reading our messages. Now, we do not want to give a detailed account of how Google works here but it is just worth mentioning that Google employs a ranking system to assess the quality of a page, called PageRank, that makes use of the link structure of the Web. The intuitive justification is that, as Brin and Pages put it, "a page can have a high PageRank if there are many pages that point to it, or if there are some pages that point to it and have a high PageRank" [Brin and Page, 1998].[3] What does that mean? It means that Google has acquired an independent status from its authors. More precisely, Google is indeed the algorithm implemented by a bunch of smart programmers, but also the aggregated choices of the millions of users who daily surf the Net. Hence the question: who or what really accesses and reads your email?

The example points to the conclusion that some technologies really exist with a moral agency that cannot be reduced to their authors or anybody else. Indeed, Google has no soul, no intentions, nothing I would call *human*. However, we may argue that Google pragmatically enhances or diminishes our capacity of acting morally. That is, it has a pragmatic impact on our moral lives.

In the example of Google, a technological artifact pragmatically (or externally) changes the moral tasks one may face. Following the distinction between pragmatic and epistemic action, I maintain that technological artifacts can also internally shape our moral performances and the capacity of seeing moral entities.

To clarify this point, consider the following case. Suppose that John has quarreled with a friend named Jane. John is very angry and thus he decides to write an email to Jane in which he expresses his profound irritation. Once finished, John re-reads the message he furiously typed and then decides that it is too nasty to send to his friend. A "sending confirmation message" pops up and he decides not to send his email. What is the cognitive meaning of John's decision? In this case, John's decision not to email Jane can be considered as a result of a manipulative activity that is mainly tacit and implicit, in which the role of the external resources (software, in this case) is crucial. The decision to write and then re-read allows him to manipulate his feelings and emotions so that new and previously unavailable information and reasons are successfully unearthed. Reasons, for instance, that make John think that his friend does not deserve to receive the words he just wrote.

I claim that this last consideration questions some assumptions related to moral agents. As already mentioned, computers have dramatically increased the moral impact of external supports over our lives. That is, the computer is a result of a massive cognitive delegation[4] and for this there may a high moral price to pay. Getting ever more complex, computers have gained a sort of moral agency [Cartesin-Stahl, 2004; Floridi and Sanders, 2004]. That is not to say that computers possess intentions, desires, emotions or some kind of moral understanding. However, computers can be viewed as surrogate agents [Johnson, 2004]. As Deborah Johnson put it, "[a]rtifacts

---

[3] The same idea is found in the Academic field: if your work is cited by many other scholars and researchers, or if it is cited by scholars and researchers who are, in turn, cited by many others, it acquires value. See for example [Bornman and Hans-Dieter, 2008].

[4] This massive delegation gave birth to what Magnani [2006a] called the "mimetic mind". That is, computers do not only mimetically reproduce certain cognitive performances (for instance, calculating), but the entire mind. In Magnani's view, this is connected with the notion of the Turing machine as a Universal one: "Computers – he wrote – are mimetic minds because they are able to mimic the mind in a kind of universal way".

are intentional insofar as they are poised to behave in a certain way". However, she adds, "[B]oth inputs from users and outputs of the artifacts can be unanticipated, unforeseen and harmful" [Johnson, 2004]. Johnson's line of debate is twofold: a) somehow computers embody our intentions; 2) but, the outputs of artifacts as well as the input of users cannot be foreseen.

Therefore, computers become to some extent autonomous. The example of Google™clearly makes the point, since we are dealing with a powerful technology that is not totally dependent on its authors. Johnson's argument is pretty consistent and I almost agree with it. However, it seems to reduce the overall moral impact of technology. My contention is that Johnson just looks at the pragmatic level of the interaction between humans and technology, but she completely discards the epistemic one. In the second example introduced above, email does not pragmatically alter the task one actually faces, that is, it does not bring John closer to his goal. But it changes the representation of the problem in such a way that he can acquire more information and thus make a better decision. What kind of agency is that? In order to provide a sound answer in the next section I will treat the notion of moral mediator.

6.2.3   Moral Mediators and External Representations

The distinction between pragmatic and epistemic action leads us to acknowledge the role of representations in problem-solving and decision-making. More precisely, I have shown with the example of email how various external artifacts (a software, for instance) can drastically change the way we cope with a decision or a problem. I may argue that computers and various technologies can be considered as external representations that alter the moral and cognitive task one faces.

More generally, I can say that we have: 1) a goal G to reach; 2) an initial state IS that is the starting point. Then we may have two or more competitive representations of the task $RT1$ and $RT2$. The representation of the problem can be viewed as the set S of mental operations, but also of actions, manipulations, inferences that we are prompted to obtain our goal. Within this framework, in the case of email, it generates an external representation that helps John to manipulate his emotions and feelings. I can represent this process with the diagram below:

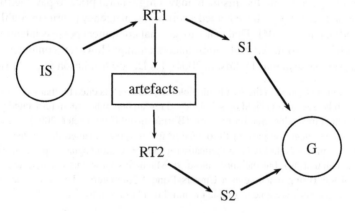

Now, John quarreled with a friend and now he is really angry. Suppose $RT1$ is the representation of the problem he has at t. $RT1$ is not adequate to make a decision, therefore, he decides to start writing an email. Now, email allows him to generate an alternative representation $RT2$ that provides him with additional resources, and thus make a better decision. According to my view, I may draw some theoretical implications. First, moral reasoning is always a mediated process. That is to say, moral tasks are always mediated by a representation[5]. Second, the representation of a task is not only a mental structure. But it can be also viewed as a step by step procedure that emerges from the interaction between humans and the environment. Thus, a representation is also something that happens outside the human mind; in this sense, I may say that a representation is something that is occurring both internally and externally [Gatti and Magnani, 2006; Knuuttila and Honkela, 2005; Wilson, 2004] (cf. section 3.2.2). Third, artifacts, tools, computers, for example, can shape, and even change the representation of a problem so as to make the solution more transparent or uncover new valuable information. Following Magnani [2007c], computers and other various artifacts can be called moral mediators. In order to shed light on this issue let me make a simple example. According to a recent survey the amount of dollars paid by US to go to war in Iraq is currently about 173 billions. This is a huge amount of money, indeed. However, if we look closer, a number does not show a lot of things. That is, it is not always simple to figure out what "173 billions" really means. Therefore, we need to compare large amounts of money to something else, for instance, to our salary, to make sense of it. Costofwar.com has tried to make sense of the enormous amount of dollars spent for the war in Iraq: it provides a very interesting representation. Let me consider the figure below:

In this case the representation provided by the website does not only consist in a twelve digit number, but it is a number that is constantly being updated live by a java script technology. Moreover, it is also compared live as well with what we could have done instead of war. The website furnishes many comparisons, for instance, with the number of children we could have insured, but also with the number of four-year scholarships which could have been provided at public schools, and so on.

---

[5] It is worth noting that this conclusion is consistent with the one proposed by Peirce [1967], who argued that we have no power of thinking without signs. For a semiotic account, see Magnani [2007b]. Cf. also Chapter 4.

This example does not add anything to the fact of the cost of the war in Iraq. However, it represents the same piece of information so that the problem we face, for instance, thinking about going to war or not, is completely changed. First of all, because we have a vivid idea about the amount of money spent. Time provides fundamental clues to make sense of if: we are being updated every second. This representation furnishes also useful comparisons so that it compel us to also think also about the rightness of that war. For instance, would not it be better to pursue different and more life-oriented policies? Was that war so necessary?

I maintain that the website uncovers and unearths certain information that otherwise would have remained invisible or unavailable for making sound judgments. More generally, the point is that without external resources, such as those in the website, we would have had to make a greater effort to get an idea of the cost of war and it would have been highly time-consuming.

Once again we have a problem-solving activity to accomplish and a problem with moral and political issues. Here I contend that the website can be considered a moral mediator, because it mediates the task, changing the representation we have of it and making the solution more transparent. More generally moral mediator refers to all those situations in which various external resources overcome the poverty and the unsatisfactory character of the moral options immediately represented or found internally. That is, a moral mediator consists the redistribution of moral effort through managing and manipulating objects and external representations.

I argue that the notion of moral mediator can help to solve some of the problems related to moral agents I introduced above. As already mentioned, I agree with Deborah Johnson [2004] who claimed computers and various technological devices are surrogate agents. On the one hand, they embody human intensions; but, on the other, they are becoming ever more autonomous as their complexity increases. Within our framework, surrogate (moral) agents are those which have a pragmatic impact over our lives. Google^{TM}is an example of this kind.

However, I argued that this view does not consider the cognitive impact of technology. That is, computers do not only replicate or replace some kind of human behavior but they also bring into existence and find room for new views and/or ideas. As moral mediators, artifacts are not only surrogate agents, but they redefine the boundaries of human moral agency. That is, they are part of it. More precisely, human moral agency is distributed and hybrid, and it is continuously shaped by the interplay between an individual and the environment.

6.3   A Case in Point: The Internet as a Moral Mediator

In this applied section I deal with the possible impact of Internet on democracy. The question is: why may the Internet challenge and help democracy? What kind of activities may the Internet mediate in order to foster our crippled democracies?

Generally speaking, I maintain that the Internet, as a moral mediator, may enhance democracy in two respects. First, the Internet allows people to confront different sources of information so that almost everyone can verify and test the information delivered by traditional media; second, it affords civic engagement and participation. More precisely, the Internet can be considered as a community builder.

6.3.1 Information as Democratic Resources

What people think, what their preferences are, become especially important in democracy. As Karl Popper [1945] argued, the appeal of democracy rests on the possibility of getting rid of those who rule without bloodshed, but through general elections. Whereas in the other forms of government those who are ruled must make a revolution to dismiss who rules: that is, the force of the best army is the necessary condition to change the government. On the contrary, in democracy information matters, not weapons, because voting is based on ideas and arguments which one can have or get. This leads to two interesting consequences: first, people can influence (and be influenced by) others' views to orient policy makers. Second, public debates and discussions are fundamental to accomplish this task. In this sense, I maintain that a deliberative version fairly represents the appeal of democracy. That is, I claim that discussing matters before making some collective decision constitutes the rationale of democracy; the more people can freely access and participate in public debates and face different opinions, the more democracy serves its purpose.

I provide two arguments to support this conclusion. The first is moral. Discussions allow people to express and debate their preferences. That is, everyone has the chance to have her/his say. Therefore, this makes people more inclined to accept the outcome of a vote, no matter what it would be, because they had the opportunity to discuss it. Moreover, the fact that people can have their say implies everyone has to provide a justification of their ideas. Those claims that cannot be reasonably supported might be discarded.

The second argument is a cognitive one. Debate reveals private information that otherwise would remain folded. Simply having a vote does not contribute to express what one thinks and, most of all, how intense one's preference is [Fearon, 1998]. This is crucial to compare different instances and solve inconsistencies. Moreover, discussions are important also for lessening "bounded" rationality. In this case, debate allows people to pool their limited capabilities through discussion [Simon, 1983].

The rationale of democracy rests on its deliberative nature and the fact that none can be excluded, however democracy does not prevent from possible damage or degeneration. As already mentioned, in democracy people's preferences acquire great importance, since people base their vote upon the information and the arguments they face and gather. That is, citizens vote for those who support, or are closest to, their own ideas. However, this is only one face of the coin when considering propaganda. Indeed, propaganda is a necessary condition to keep democracy working. As Bernays put it, "a desire of a specific reform, however widespread, cannot be translated into action until it is made articulate" [Bernays, 2005, p. 57]. As a matter of fact, a desire for a certain policy does not come up to the citizens' minds simultaneously [Lippmann, 1997, p. 155]: public opinion must be focused and organized.

However, citizens' preferences can be easily manipulated and even manufactured. As Chomsky put it, in democracy the government cannot control people by force, but "it can control people's minds" [Chomsky, 2002, p. 2001]. Therefore, the way people can access information, how they build their preferences up, is a key issue to prevent democracy from degeneration.

Now, my claim is that the Internet drastically changes the way people can get and share information. First of all, traditional media (especially those related to the news) can be easily manipulated and controlled by the political power who often boosts its agenda by biased, or even bribed, columnists or editors [Furedi, 2002]. In contrast, the Internet (and the Net), as an unstructured and ever growing information space, seems to reduce the overall power of government to control citizens [Simon, 2002]. The Internet and the Web in particular are searching environments in which people are enabled to search for whatever they want without any kind of filtering. They can access various sources of information and exploit social sources of information such as forums, chat rooms and blogs. In this sense, the Internet dramatically changes the task people face, when they deal with political issues.

6.3.2 The Internet as a Community Builder

The Internet may enhance democracy in another respect that is related to the problem of political participation and civic engagement. As already mentioned, the rationale of democracy not only concerns voting, but also debating and discussing. However, discussing and debating presuppose that people are truly engaged in all those activities that involve public life. As Putnam [2000] suggested, it is more likely that democracy spreads, when the so called connective tissue of the society is highly developed. The more people are separated from each other, the more the political engagement drastically decreases.

Now, the claim that the Internet allows people to search for whatever they want is well-founded, but it is not the whole deal. As maintained by Meikle [2002], the Internet is not only a medium of consumption, but also of intercreativity. For instance, reading a newspaper is a kind of activity that presupposes a one-way communication flow, so to say. I can read what an editor writes, but she cannot read what I would like to write to her. In this sense, people are primarily information consumers. On the contrary, the new technologies that belong to the so called "Internet Galaxy" [Castells, 2001] make intercreativity possible. By the term intercreativity, we mean something more than simple interactivity. In order to define what intercreativity is, I have to introduce some important distinctions.

For example, several on-line newspapers allow people to select what they want to read or receive in their e-mailbox. Moreover, in many cases, one can post a comment on a given article. However, almost always the options available to the user is limited and already selected by the editor. This is the kind of interactivity exhibited by a jukebox. Second, one can post some comments about a certain story which has been published, but he cannot modify it. These are two examples among others that fairly represent kinds of interactivity.

On the contrary, by the term intercreativity I simply refer to something that is created by a truly two- way communication flow, in which everyone can contribute to producing, choosing, and modifying a given document (an article or the course of an open discussion). For instance, an email exchange with a friend or a forum are examples of this kind.

Now, the fact that the Internet exhibits this kind of intercreativity can play a crucial role in enhancing civic engagement. As mentioned above, the more people are separated from each other, the more the political engagement drastically decreases. Now, the point is that the Internet provides citizens with new possibilities that drastically change the way people can reach each other. That is, citizens cease to be information consumers and become participants, that is a sort of necessary condition to keep democracy working. There are plenty of examples where new political strategies of civic engagement are brought about. No matter where they are, people can share information, make common cause, and jointly advance their mutual political or other agendas [Simon, 2002]. Mailing list, newsletters, forums, on-line conference tools, contribute to boost civic engagement. Besides, it is worth noting that also the idea of open publishing promotes those values that are very close to democracy, such as freedom of speech, and so forth.

6.4 Concluding Remarks

In this chapter I have tried to outline – what I called – a morality of everyday things. I aimed at providing an alternative framework in which various technologies are not considered as mere instruments, but as part of our cognitive and moral system. Taking advantage of the notion of distributed morality introduced by Magnani, I pointed out that computers do not only pragmatically interact with humans, (and eventually create new ethical problems), but they also actively shape the way humans solve ethical problems or generate new ideas.

References

[Adolphs, 2001] Adolphs, R.: The neurobiology of social cognition. Current Opinion in Neurobiology 1, 231–239 (2001)

[Adolphs, 2006] Adolphs, R.: How do we know the minds of others? domain specificity, simulation and enactive social cognition. Brain Research 1079, 25–35 (2006)

[Alvard, 2003] Alvard, M.S.: The adaptive nature of culture. Evolutionary Anthropology 12, 136–149 (2003)

[Anderson et al., 2005] Anderson, M., Anderson, S.L., Armen, C.: Machine ethics. papers from the aaai fall symposium, Technical Report FS0506. AAAI Press, Menlo Park, CA (2005)

[Anderson, 2006] Anderson, M.H.: How can we know what we think until we see what we said?: A citation and citation context analysis of Karl Weick The Social Psychology of Organizing. Organization Studies 27(11), 1675–1692 (2006)

[Aunger, 2002] Aunger, R.: The Eletric Meme. A New Theory of How We Think. The Free Press, New York (2002)

[Axelrod, 1984] Axelrod, R.: The Evolution of Cooperation. Basic Books, New York (1984)

[Bandura, 1999] Bandura, A.: Moral disengagement in the perpetration of inhumanities. Personality and Social Psychology Review 3, 193–209 (1999)

[Bardone and Secchi, 2009] Bardone, E., Secchi, D.: Distributed cognition: A research agenda for management. In: Rahim, M.A. (ed.) Current Topics Management, vol. 14, pp. 183–207. Transaction Publishers, New Brunswick (2009)

[Batelle, 2005] Batelle, J.: The Search: How Google and Its Rivals Rewrote the Rules of Business and Transformed Our Culture. Nicholas Brealey Publishing, New York (2005)

[Baumeister et al., 2004] Baumeister, R.F., Zhang, L., Vohs, K.D.: Gossip as cultural learning. Review of General Psychology 8, 111–121 (2004)

[Beach, 1997] Beach, L.R.: The Psychology of Decision Making. People in Organization. Sage, Thousand Oaks, CA (1997)

[Beach, 1998] Beach, L.R. (ed.): Image theory. Theoretical and Empirical Foundations. Lawrence Erlbaum Associates, Mahwah, NJ (1998)

[Becchio et al., 2007] Becchio, C., Pierno, A.C., Mari, M., Lusher, D., Castello, U.: Motor contagion from eye gaze. the case of autism. Brain 130, 2401–2411 (2007)

[Ben-Séev, 1994] Ben-Séev, A.: The vindication of gossip. In: Goodman, R.F., Ben-Ze'ev, A. (eds.) Good Gossip, pp. 11–24. University Press of Kansas, Lawrence (1994)

[Bernays, 2005] Bernays, E.: Propaganda. IG Publishing, Brooklin (2005) (original edition published in 1928)

[Bernstein et al., 2008] Bernstein, M., van Kleek, M., Karger, D., Schraefel, M.C.: Information scraps: How and why information eludes our personal information management tools. ACM Transactions on Information Systems 26(4), 1–46 (2008)

[Bertolotti and Magnani, 2010] Bertolotti, T., Magnani, L.: The role of agency detection in the invention of supernatural beings: an abductive approach. In: Magnani, L., Carnielli, W., Pizzi, C. (eds.) Model-Based Reasoning in Science and Technology. Studies in Computational Intelligence, vol. 314, pp. 239–262. Springer, Heidelberg (2010)

[Bingham, 1999] Bingham, P.M.: Human uniqueness: a general theory. The Quarterly Review of Biology 74(2), 133–169 (1999)

[Bingham, 2000] Bingham, P.M.: Human evolution and human history: a complete theory. Evolutionary Anthropology 9(6), 248–257 (2000)

[Boehm, 1999] Boehm, C.: Hierarchy in the Forest. Harvard University Press, Cambridge, MA (1999)

[Bohlen and Frei, 2009] Bohlen, M., Frei, H.: Ambient Intelligence in the city. Overview and new perspectives. In: Nakashima, H., Aghajan, H., Augusto, J.C. (eds.) Handbook of Ambient Intelligence and Smart Environments, pp. 911–938. Springer, Heidelberg (2009)

[Bornman and Hans-Dieter, 2008] Bornman, L., Hans-Dieter, D.: What do citation counts measure? A review of studies citing behavior. Journal of Documentation 64, 45–80 (2008)

[Bornmann and Daniel, 2006] Bornmann, L., Daniel, H.-D.: What do citation counts measure? A review of studies on citing behavior. Journal of Documentation 64(1), 45–80 (2006)

[Brin and Page, 1998] Brin, S., Page, L.: The anatomy of a large-scale hypertextual web search engine (1998),
http://www-db.stanford.edu/backrub/google.html

[Brown and Kagel, 2009] Brown, A.L., Kagel, J.K.: Behavior in a simplified stock market: the status quo bias, the disposition effect and the ostrich effect. Annals of Finance 5, 1–14 (2009)

[Bruderer and Singh, 1996] Bruderer, E., Singh, J.V.: Organizational evolution, learning, and selection: A genetic-algorithm-based model. Academy of Management Journal 39(5), 1322–1349 (1996)

[Brunswik, 1943] Brunswik, E.: Oranismic achievement and environmental probability. Psychological Review 50, 255–272 (1943)

[Brunswik, 1952] Brunswik, E.: The Conceptual Framework of Psychology. University of Chicago Press, Chicago (1952)

[Brunswik, 1955] Brunswik, E.: Representative design and probabilistic theory in a functional psychology. Psychological Review 62, 193–217 (1955)

[Buccino et al., 2009] Buccino, G., Sato, M., Cattaneo, L., Roda, F., Riggio, L.: Broken affordances, broken objects: A TMS study. Neuropsychologia 47, 3074–3078 (2009)

[Buchholz and Rosenthal, 2002] Buchholz, R.A., Rosenthal, S.B.: Technology and business. rethinking the moral dilemma. Journal of Business Ethics 41, 45–50 (2002)

[Bynum and Rogerson, 2004] Bynum, T.W., Rogerson, S. (eds.): Computer Ethics and Professional Responsibility. Blackwell, Malden, MA (2004)

[Byrne and Whiten, 1997] Byrne, R.W., Whiten, A.: Machiavellian intelligence. In: Whiten, A., Byrne, R.W. (eds.) Machiavellian Intelligence II, pp. 1–23. Cambridge University Press, Cambridge (1997)

[Calvi and Magnani, 2002] Calvi, L., Magnani, L.: Mediated knowledge on the web: The role of cognitive mediators in design. In: Proceedings of the ECCE 11, Cognition, Culture and Design, Catania, Italy (2002)

[Carey et al., 1996] Carey, D.P., Harvey, M., Milner, A.D.: Visuomotor sensitivity for shape and orientation in a patient with visual form agnosia. Neuropsychologia 34(5), 329–337 (1996)

[Carnap, 1947] Carnap, R.: On the application of inductive logic. Philosophy and Phenonmenlogical Research 8, 133–148 (1947)

[Carnielli, 2010] Carnielli, W.: On a theoretical analysis of deceiving: How to resist a bullshit attack. In: Magnani, L., Carnielli, W., Pizzi, C. (eds.) Model-Based Reasoning in Science and Technology. Abduction, Logic and Computational Discovery. Springer, Berlin (2010) (in press)

[Carson, 2010] Carson, T.L.: Lying and Deception. Theory and Practice. Oxford University Press, Oxford (2010)

[Cartesin-Stahl, 2004] Cartesin-Stahl, C.B.: Information, ethics, and computers. the problem of autonomous moral agent. Minds and Machines 14, 67–83 (2004)

[Case and Higgins, 2000] Case, D.O., Higgins, G.M.: How can we investigate citation behavior? A study of reasons for citing literature in communication. Journal of the American Society for Information Science 51(7), 635–645 (2000)

[Castelli et al., 2000] Castelli, F., Happe, F., Frith, U., Frith, C.: Movement and mind: a functional imaging study of perceptios and interpretation of complex intentional movement patterns. Neuroimage 12, 314–325 (2000)

[Castells, 2001] Castells, M.: The Internet Galaxy: Reflections on the Internet, Business and Society. Oxford University Press, Oxford (2001)

[Castro et al., 2004] Castro, L., Toro, M.A., Ayala, F.J.: The evolution of culture: from primate social learning to human culture. Proceedings of the National Academy of Sciences of the United States of America 101(27), 10235–10240 (2004)

[Chawner, 2005] Chawner, B.: Free/open source software: New opportunities, new challenges (2005),
http://www.vala.org.au/vala2004/2004pdfs/33Chawn.PDF

[Chemero, 2003] Chemero, A.: An outline of a theory of affordances. Ecological Psychology 15(2), 181–195 (2003)

[Chomsky, 2002] Chomsky, N.: Understanding Power. New Press, New York (2002)

[Clancey, 1997] Clancey, W.J.: Situated Cognition: on Human Knowledge and Computer Representations. Cambridge University Press, Cambridge (1997)

[Clark and Chalmers, 1998] Clark, A., Chalmers, D.J.: The extended mind. Analysis 58, 10–23 (1998)

[Clark, 1997] Clark, A.: Being There: Putting Brain, Body, and World Together Again. The MIT Press, Cambridge, MA (1997)

[Clark, 2003] Clark, A.: Natural-Born Cyborgs. Minds, Technologies and the Future of Human Intelligence. Oxford University Press, Oxford (2003)

[Clark, 2006] Clark, A.: Language, embodiment and the cognitive niche. Trends in Cognitive Science 10(8), 370–374 (2006)

[Clark, 2008] Clark, A.: Supersizing the Mind. Embodiment, Action and Cognitive Extension. Oxford University Press, Oxford (2008)

[Conlin, 2009] Conlin, J.A.: Getting around: making fast and frugal navigation decisions. In: Markus Raab, M., Johnson, J.G., Heekeren, H.R. (eds.) Mind and Motion: The Bidirectional Link between Thought and Action, vol. 174, pp. 109–117. Elsevier, Amsterdam (2009)

[Conlisk, 1996] Conlisk, J.: Why bounded rationality? Journal of Economic Literature 34, 669–700 (1996)

[Cook and Das, 2007] Cook, J.D., Das, S.K.: How smart are our environments? An updated look at the state of the art. Pervasive and Mobile Computing 3(2), 53–73 (2007)

[Cook et al., 2009] Cook, D.J., Augusto, J.C., Vikramaditya, R.J.: Ambient intelligence: Technologies, applications, and opportunities. Pervasive and Mobile Computing 5, 277–298 (2009)

[Curley and Keverne, 2005] Curley, J.P., Keverne, E.B.: Genes, brains and mammalian social bonds. Trends in Cognitive Science 20(10), 561–566 (2005)

[Cyert and March, 1963] Cyert, R.M., March, J.G.: A Behavioral Theory of the Firm. Prentice-Hall, Englewood Cliffs (1963)

[Damasio, 1999] Damasio, A.R.: The Feeling of What Happens. Harcourt Brace, New York (1999)

[Dautenhan, 2001] Dautenhan, K.: The narrative intelligence hypothesis: in search of the transactional format of narrativs in humans and other social animals. In: Beynon, M., Nehaniv, C.L., Dautenhan, K. (eds.) CT 2001. LNCS (LNAI), vol. 2117, pp. 248–266. Springer, Berlin (2001)

[Dawkins, 2004] Dawkins, R.: Extended phenotype - but not extended. A reply to Laland, Turner and Jablonka. Biology and Philosophy 19, 377–397 (2004)

[De George, 1999] De George, R.T.: Business ethics. Prentice Hall, Upper Saddle River (1999)

[de Leon, 2002] de Leon, D.: Cognitive task transformations. Cognitive Systems Research 3, 349–359 (2002)

[Decety and Grèzes, 2006] Decety, J., Grèzes, J.: The power of simulation: imagining one's own and other behavior. Brain in Research 1079, 4–14 (2006)

[D'Errico and Cacho, 1994] D'Errico, F., Cacho, C.: Notion versus decoration in the upper Paleolithic: a case study from Tossal de la Roca, Alicante, Spain. Journal of Archaeological Science 21, 185–200 (1994)

[Dessalles, 2000] Dessalles, J.-L.: Language and hominid politics. In: Hurford, J.R., Knight, C., Suddert-Kennedy, M. (eds.) The Evolutionary Emergence of Language: Social Function and the Origin of Linguistic Form, pp. 62–79. Cambridge University Press, Cambridge (2000)

[Donald, 2001] Donald, M.: A Mind So Rare. The Evolution of Human Consciousness. Norton, London (2001)

[Dunbar, 1996] Dunbar, R.I.M.: Grooming, Gossip, and the Evolution of Language. Harvard University Press, Cambridge, MA (1996)

[Dunbar, 1998] Dunbar, R.I.M.: The social brain hypothesis. Evolutionary Anthropology 6, 178–190 (1998)

[Dunbar, 2004] Dunbar, R.I.M.: Gossip in an evolutionary perspective. Review of General Psychology 8, 100–110 (2004)

[Eashtem and Easthem, 1991] Eashtem, M., Easthem, A.: Paleolithic parietal art and its topographic context. Proceeding of the Prehistoric Society 51, 115–128 (1991)

[Efferson et al., 2008] Efferson, C., Lalive, R., Richerson, P.J., McElreath, R., Lubell, M.: Conformists and mavericks: the empirics of frequency-dependent cultural transmission. Evolution and Human Behavior 29, 56–64 (2008)

[Ellis, 1995] Ellis, R.D.: Questioning Consciousness: the Interplay of Imagery, Cognition and Emotion in the Human Brain. John Benjamins, Amsterdam (1995)

[Eng et al., 2005] Eng, K., Douglas, R.J., Verschure, P.F.M.J.: An interactive space that learns to influence human behavior. IEEE Transactions on Systems, Man and Cybernetics, Part A 35(1), 66–77 (2005)

[Etzioni, 1988] Etzioni, A.: The moral dimension. Toward a new economics. The Free Press, New York (1988)

[Evans, 2002] Evans, J.S.B.T.: Logic and human reasoning: an assessment of deduction. Psychological Bulletin 128(8), 978–996 (2002)

[Fearon, 1998] Fearon, F.: Deliberation as discussion. In: Elster, J. (ed.) Deliberative Democracy, pp. 123–140. Cambridge University Press, Cambridge (1998)

[Field, 2008] Field, A.J.: Why multilevel selection matters. Journal of Bioeconomics 10(3), 203–238 (2008)

[Fletcher and Zwick, 2004] Fletcher, J.A., Zwick, M.: Strong altruism can evolve in randomly formed groups. Journal of Theoretical Biology 228, 303–313 (2004)

[Flinn et al., 2005] Flinn, M., Geary, D., Ward, C.: Ecological dominance, social competition, and coalitionary arms races. Why humans evolved extraordinary intelligence. Evolution and Human Behavior 26(1), 10–46 (2005)

[Floridi and Sanders, 2004] Floridi, L., Sanders, J.W.: On the morality of artificial agents. Minds and Machines 14, 349–379 (2004)

[Floridi, 1999] Floridi, L.: Philosophy and Computing. Routledge, London, New York (1999)

[Foss, 2003] Foss, N.J.: Bounded rationality in the economics of organizations: 'much cited and little used'. Journal of Economic Psychology 24, 245–264 (2003)

[Fotopoulou et al., 2007] Fotopoulou, A., Conway, M.A., Solms, M.: Confabulation: Motivated reality monitoring. Neuropsychologia 45, 2180–2190 (2007)

[Frank, 1988] Frank, R.H.: Passions within reason. W.W. Norton, New York (1988)

[Frank, 2004] Frank, R.H.: What Price the Moral High Ground? Princeton University Press, Princeton (2004)

[Frankfurt, 2005] Frankfurt, H.: On Bullshit. Princeton University Press, New York (2005)

[Frischen et al., 2007] Frischen, A., Bayliss, A.P., Tipper, S.P.: Gaze-cueing of attention: Visual attention, social cognition and individual differences. Psychological Bulletin 133(4), 694–724 (2007)

[Frischen et al., 2009] Frischen, A., Loach, D., Tipper, S.P.: Seeing the world through another person's eyes: Simulating selective attention via action observation. Cognition 111(2), 212–218 (2009)

[Furedi, 2002] Furedi, F.: Culture of Fear. Continuum, London (2002)

[Gabbay and Woods, 2001] Gabbay, D., Woods, J.: The new logic. Logic Journal of the IGPL 9(2), 141–174 (2001)

[Gabbay and Woods, 2005] Gabbay, D.M., Woods, J.: The Reach of Abduction. A Practical Logic of Cognitive Systems, vol. 2. North-Holland, Amsterdam (2005)

[Gatti and Magnani, 2006] Gatti, A., Magnani, L.: On the representational role of the environment and on the cognitive nature of manipulations. In: Magnani, L., Dossena, R. (eds.) Computing, Philosophy and Cognition, pp. 227–242. College Publications, London (2006)

[Gaver, 1991] Gaver, W.W.: Technology affordances. In: CHI 1991 Conference Proceedings, pp. 79–84 (1991)

[Gazzaniga, 2005] Gazzaniga, M.: The Ethical Brain. Dana Press, New York, Washington (2005)

[Gebhard et al., 2004] Gebhard, U., Nevers, P., Billmann-Mahecha, E.: Moralizing trees: anthropomorphism and identity in children's relationships to nature. In: Clayton, S., Opotow, S. (eds.) Identity and the Natural Environment, pp. 91–112. The MIT Press, Cambridge, MA (2004)

[Gibson and Pick, 2000] Gibson, E.J., Pick, A.D.: An Ecological Approach to Perceptual Learning and Development. Oxford University Press, Oxford (2000)

[Gibson, 1951] Gibson, J.J.: What is a form? Psychological Review 58, 403–413 (1951)

[Gibson, 1979] Gibson, J.J.: The Ecological Approach to Visual Perception. Houghton Mifflin, Boston, MA (1979)

[Gigerenzer and Brighton, 2009] Gigerenzer, G., Brighton, H.: Homo heuristicus: Why biased minds make better inferences. Topics in Cognitive Science 1, 107–143 (2009)

[Gigerenzer and Goldstein, 1996] Gigerenzer, G., Goldstein, D.G.: Reasoning the fast and frugal way: Models of bounded rationality. Psychological Review 103, 650–669 (1996)

[Gigerenzer and Selten, 2001] Gigerenzer, G., Selten, R.: Bounded Rationality: The Adaptive Toolbox. Cambridge University Press, Cambridge (2001)

[Gigerenzer, 2000] Gigerenzer, G.: Adaptive thinking: Rationality in the Real World. Oxford University Press, Oxford (2000)

[Godfrey-Smith, 1998] Godfrey-Smith, P.: Complexity and the Function of Mind in Nature. Cambridge University Press, Cambridge (1998)

[Godrey-Smith, 2002] Godrey-Smith, P.: Environmental complexity and the evolution of cognition. In: Sternberg, R., Kaufman, K. (eds.) The Evolution of Intelligence, pp. 233–249. Lawrence Erlbaum Associates, Mawhah, NJ (2002)

[Gooding, 1994] Gooding, D.: Experiment and the Making of Meaning. Human Agency in Scientific Observation and Experiment. The MIT Press, Cambridge, MA (1994)

[Gooding, 2004] Gooding, D.: Seeing the forest for the trees: Visualization, cognition and scientific inference. In: Gorman, M., Gooding, D., Tweney, R., Kincannon, A. (eds.) Scientific and Technological Thinking, pp. 173–217. Lawrance Erlaum Publishers, Mahwah, N.J. (2004)

[Goodwin, 1998] Goodwin, M.: Cyber Rights. Defending Free Speech in the Digital Age. Random House, Toronto (1998)

[Greene, 2003] Greene, J.: From neural is to moral ought: what are the moral implications of neuroscientific moral psychology? Nature Review Neuroscience 4(10), 846–849 (2003)

[Greeno, 1994] Greeno, J.G.: Gibson's affordances. Psychological Review 101(2), 336–342 (1994)

[Grice, 1975] Grice, H.P.: Logic and conversation. In: Sternberg, R., Kaufman, K. (eds.) Syntax and Semantics 3: Speech Acts, pp. 41–58. Academic Press, New York (1975)

[Gutwirth, 2009] Gutwirth, S.: Beyond identity. In: IDIS (2009) (in press)

[Hammond and Steward, 2001] Hammond, K.R., Steward, T.R. (eds.): The Essential Brunswik. Beginnings, Explications, Applications. Oxford University Press, Oxford (2001)

[Hammond et al., 1987] Hammond, K.R., Hamm, R.M., Grassia, J., Pearson, T.: Direct comparison of intuitive and analytical cognition in expert judgment. IEEE Transactions on Systems, Man and Cybernetics SMC-17, 753–770 (1987)

[Hanoch, 2002] Hanoch, Y.: Neither an angel nor an ant: Emotion as an aid to bounded rationality. Journal of Economic Psychology 23, 1–25 (2002)

[Hansen, 2002] Hansen, H.H.: The straw thing of fallacy theory: the standard definition of fallacy. Argumentation 16(2), 133–155 (2002)

[Harris, 2004] Harris, R.: Integrationism, language, mind and world. Language Sciences 26, 727–739 (2004)

[Harris, 2009] Harris, R.: Rationality and the Literate Mind. Routledge, London (2009)

[Hempel, 1966] Hempel, C.G.: Philosophy of Natural Science. Prentice-Hall, Englewood Cliffs, NJ (1966)

[Heschong, 2002] Heschong, L.: Daylighting and human performance. ASHRAE Journal 44(6), 65–67 (2002)

[Hildebrandt, 2008a] Hildebrandt, M.: Ambient intelligence, criminal liability and democracy. Criminal Law and Philosophy 2(2), 163–180 (2008)

[Hildebrandt, 2008b] Hildebrandt, M.: A vision of ambient law. In: Brownsword, R., Yeung, K. (eds.) Regulating Technologies: Legal Futures, Regulatory Frames and Technological Fixes, pp. 175–191. Hart Publishing, Oxford (2008)

[Hill and Dunbar, 2003] Hill, R.A., Dunbar, R.I.M.: Social network size in humans. Human Nature 14(1), 53–72 (2003)

[Himanen, 2001] Himanen, P.: The Hacker Ethic. A Radical Approach to Philosophy of Business. Random House, New York (2001)

[Himmelbach and Karnath, 2005] Himmelbach, M., Karnath, H.-O.: Dorsal and ventral stream interaction: contributions from optic ataxia. Journal of Cognitive Neuroscience 17(4), 632–640 (2005)

[Himmelbach et al., 2006] Himmelbach, M., Karnath, H.-O., Perenin, M.T., Franz, V.H., Stockmeier, K.: A general deficit of the 'automatic pilot' with posterior parietal cortex lesions? Neuropsychologia 44(13), 2749–2756 (2006)

[Hintikka, 2004] Hintikka, J.: A fallacious fallacy? Synthese 140, 25–35 (2004)

[Hirstein, 2005] Hirstein, W.: Brain Fiction. Self-Deception and the Riddle of Confabulation. The MIT Press, Cambridge, MA (2005)

[Hirstein, 2009] Hirstein, W.: Introduction. what is confabulation? In: Hirstein, W. (ed.) Confabulation: Views from Neuroscience, Psychiatry, Psychology and Philosophy, pp. 1–12. Oxford University Press, Oxford (2009)

[Hoffman, 1998] Hoffman, D.D.: Visual Intelligence: How We Create What We See. Norton, New York (1998)

[Hollan et al., 2000] Hollan, J., Hutchins, E., Kirsh, D.: Distributed cognition: Toward a new foundation for human-computer interaction research (2000),
http://hci.ucsd.edu/lab/publications.htm

[Holt, 2009] Holt, R.D.: Bringing the hutchinsonian niche into the 21st century: Ecological and evolutionary perspectives. PNAS 106, 19659–19665 (2009)

[Humphrey, 1976] Humphrey, N.: The social functions of intellect. In: Bateson, P.P.G., Hinde, R.A. (eds.) Growing Points in Ethology, pp. 303–317. Cambridge University Press, Cambridge (1976)

[Hutchins, 1995] Hutchins, E.: Cognition in the Wild. The MIT Press, Cambridge, MA (1995)

[Hye, 2007] Hye, P.J.: A design study of pedestrian space as an interactive space. presented at IASDR 2007 (2007)

[Ichinose and Arita, 2008] Ichinose, G., Arita, T.: The role of migration and founder effect for the evolution of cooperation in a multilevel selection context. Ecological Modeling 210, 221–230 (2008)

[Johnson, 1993] Johnson, M.: Moral Imagination. Implications of Cognitive Science in Ethics. The Chicago University Press, Chicago, IL (1993)

[Johnson, 1994] Johnson, D.G.: Computer Ethics. Prentice Hall, Englewood Cliffs, NJ (1994)

[Johnson, 2004] Johnson, D.G.: Integrating ethics and technology. In: European Conference Computing and Philosophy E-CAP 2004, Pavia, Italy, June 2-5 (2004) (abstract)

[Jones, 1991] Jones, T.: Ethical decision-making by individuals in organizations: An issue-contingent model. The Academy of Management Review 16(2), 366–395 (1991)

[Joyce, 2006] Joyce, R.: The Evolution of Morality. The MIT Press, Cambridge, MA (2006)

[Kahneman et al., 1990] Kahneman, D., Knetsch, J.L., Thaler, R.H.: Experimental tests of the endowment effect and the coase theorem. Journal of Political Economy 98, 1325–1348 (1990)

[Kahneman, 2003] Kahneman, D.: A perspective on judgement and choice. mapping bounded rationality. American Psychologist 58(9), 697–720 (2003)

[Karlsson et al., 2009] Karlsson, N., Loewenstein, G., Seppi, D.: The ostrich effect: Selective attention to information. Journal Risk and Uncertainty 38, 95–115 (2009)

[Keen, 2007] Keen, A.: The Cult of Amateur. How Today's Internet is Killing Our Culture and Assaulting Our Economy. Nicholas Brealey Publishing, London (2007)

[Khalil, 2004] Khalil, E.L.: What is altruism? Journal of Economic Psychology 25, 97–123 (2004)

[Kirlik, 2001] Kirlik, A.: On gibson's review of brunswik. In: Hammond, K.R., Steward, T.R. (eds.) The Essential Brunswik. Beginnings, Explications, Applications, pp. 238–242. Oxford University Press, Oxford (2001)

[Kirsh and Maglio, 1994] Kirsh, D., Maglio, P.: On distinguishing epistemic from pragmatic action. Cognitive Science 18, 513–549 (1994)

[Kirsh, 1999] Kirsh, D.: Distributed cognition, coordination and environment design. In: Proceedings of the European Conference on Cognitive Science (1999)

[Kirsh, 2004] Kirsh, D.: Metacognition, distributed cognition and visual design. In: Gardinfors, P., Johanson, P. (eds.) Cognition, Education and Communication Technology. Lawrence Erlbaum Associates, Mahwah, N.J (2004)

[Klucharev et al., 2009] Klucharev, V., Hytonen, K., Rijpkema, M., Smidts, A., Fernandez, G.: Reinforcement learning predicts social conformity. Neuron 61, 140–151 (2009)

[Knuuttila and Honkela, 2005] Knuuttila, T., Honkela, T.: Questioning external and internal representation: the case of scientific models. In: Magnani, L., Dossena, R. (eds.) Computing, Philosophy and Cognition, London, pp. 209–226. College Pubblications (2005)

[Krechmer, 2005] Krechmer, K.: Open standards requirements. In: Proceedings of the 38th Annual Hawaii International Conference on System Sciences (2005)

[Kunda, 1999] Kunda, F.: Social Cognition. The MIT Press, Cambridge, MA (1999)

[Lahti and Weinstein, 2005] Lahti, D., Weinstein, B.S.: The better angels of our nature: group stability and the evolution of moral tension. Evolution and Human Behavior 26(1), 47–63 (2005)

[Laland et al., 2000] Laland, K.N., Odling-Smee, J., Feldman, M.W.: Niche construction, biological evolution and cultural change. Behavioral and Brain Sciences 23(1), 131–175 (2000)

[Laland et al., 2001] Laland, K.N., Odling-Smee, F.J., Feldman, M.W.: Cultural niche construction and human evolution. Journal of Evolutionary Biology 14, 22–33 (2001)

[Laland et al., 2005] Laland, K.N., Odling-Smee, J., Feldman, M.W.: On the breath and significance of niche construction: a reply to grittiths, okasha and sterelny. Biology and Philosophy 20, 37–55 (2005)

[Leibenstein, 1950] Leibenstein, H.: Bandwagon, Snob, and Veblen Effects in the theory of consumers' demand. The Quarterly Journal of Economics 64(2), 183–207 (1950)

[Lewis-Williams, 2002] Lewis-Williams, D.: The Mind in the Cave. Thames and Hudson, London (2002)

[Lewontin and Hubby, 1985] Lewontin, R.C., Hubby, J.L.: Citation classic. Current Contents/Life Science 43, 16 (1985)

[Lipman, 1995] Lipman, B.R.: Information processing and bounded rationality. The Canadian Journal of Economics 28(1), 42–67 (1995)

[Lippmann, 1997] Lippmann, W.: Public Opinion. Free Press, London (1997) (original edition published in 1921)

[Logan, 2006] Logan, R.K.: The extended mind model of the origin of language and culture. In: Gontier, N., Van Bendegem, J.P., Aerts, D. (eds.) Evolutionary Epistemology, Language and Culture, pp. 149–167. Springer, Berlin (2006)

[Love, 2004] Love, N.: Cognition and the language myth. Language Sciences 26, 525–544 (2004)

[Magnani and Bardone, 2006] Magnani, L., Bardone, E.: Designing human interfaces. the role of abduction. In: Magnani, L., Dossena, R. (eds.) Computing, Philosophy and Cognition, pp. 131–146. College Publications, London (2006)

[Magnani and Bardone, 2008] Magnani, L., Bardone, E.: Sharing representations and creating chances through cognitive niche construction. The role of affordances and abduction. In: Iwata, S., Oshawa, Y., Tsumoto, S., Zhong, N., Shi, Y., Magnani, L. (eds.) Communications and Discoveries from Multidisciplinary Data, pp. 3–40. Springer, Berlin (2008)

[Magnani, 1992a] Magnani, L.: Abductive reasoning: philosophical and educational perspectives in medicine. In: Evans, D.A., Patel, V.L. (eds.) Advanced Models of Cognition for Medical Training and Practice, Berlin, pp. 21–41. Springer, Heidelberg (1992)

[Magnani, 1992b] Magnani, L.: Abductive reasoning: philosophical and educational perspectives in medicine. In: Evans, D.A., Patel, V.L. (eds.) Advanced Models of Cognition for Medical Training and Practice, pp. 21–41. Springer, Berlin (1992)

[Magnani, 2001] Magnani, L.: Abduction, Reason, and Science. Processes of Discovery and Explanation. Kluwer Academic/Plenum Publishers, New York (2001)

[Magnani, 2005] Magnani, L.: Chance discovery and the disembodiment of mind. In: Oehlmann, R., Abe, A., Ohsawa, Y. (eds.) Proceedings of the Workshop on Chance Discovery: from Data Interaction to Scenario Creation, International Conference on Machine Learning (ICML 2005), pp. 53–59 (2005)

[Magnani, 2006a] Magnani, L.: Mimetic minds. Meaning formation through epistemic mediators and external representations. In: Loula, A., Gudwin, R., Queiroz, J. (eds.) Artificial Cognition Systems, pp. 327–357. Idea Group Publishers, Hershey, PA (2006)

[Magnani, 2006b] Magnani, L.: Prefiguring ethical chances: the role of moral mediators. In: Oshawa, Y., Tsumoto, S. (eds.) Chance Discoveries in Real World Decision Making: Data-based Interaction of Human and Artificial Intelligence, pp. 205–229. Springer, Berlin (2006)

[Magnani, 2007a] Magnani, L.: Abduction and cognition in human and logical agents. In: Artemov, S., Barringer, H., Garcez, A., Lamb, L., Woods, J. (eds.) We Will Show Them: Essays in Honour of Dov Gabbay, London, vol. II, pp. 225–258. College Publications, London (2007)

[Magnani, 2007b] Magnani, L.: Semiotic brains and artificial minds. How brains make up material cognitive systems. In: Gudwin, R., Queiroz, J. (eds.) Semiotics and Intelligent Systems Development, pp. 1–41. Idea Group Inc., Hershey, PA (2007)

[Magnani, 2007c] Magnani, L.: Morality in a Technological World. Knowledge as Duty. Cambridge University Press, Cambridge (2007)

[Magnani, 2009] Magnani, L.: Abductive Cognition. The Epistemological and Eco-Cognitive Dimensions of Hypothetical Reasoning, vol. 3. Springer, Heidelberg (2009)

[Magnani, 2011] Magnani, L.: Understanding Violence. In: Morality, Religion, and Violence Intertwined: a Philosophical Stance. Springer, Heidelberg (2011)

[Maier and Kempter, 2009] Maier, E., Kempter, G.: Aladin - a magic lamp for the elderly? In: Hideyuki Nakashima, H., Aghajan, H., Augusto, J.C. (eds.) Handbook of Ambient Intelligence and Smart Environments, pp. 1201–1227. Springer, Heidelberg (2009)

[March, 1978] March, J.G.: Bounded rationality, ambiguity and the engineering of choice. Bell Journal of Economics 9, 587–608 (1978)

[Marcus, 2004] Marcus, G.: The Birth of the Mind. How a Tiny Number of Genes Creates the Complexity of Human Thought. Basic Books, New York (2004)

[Marr, 1982] Marr, D.: Vision. Freeman, San Francisco, CA (1982)

[Maynard-Smith and Szathmary, 1995] Maynard-Smith, J., Szathmary, E.: The Major Transitions in Evolution. Freeman, Oxford (1995)

[Maynard Smith, 1987] Maynard Smith, J.: How to model evolution. In: Dupre, J. (ed.) The Latest on the Best: Essays on Evolution and Optimality, pp. 119–131. The MIT Press, Cambridge, MA (1987)

[McGrenere and Ho, 2000] McGrenere, J., Ho, W.: Affordances: clarifying and evolving a concept. In: Proceedings of Graphics Interface, Montreal, Quebec, Canada, May 15-17, pp. 179–186 (2000)

[Meikle, 2002] Meikle, G.: Future Active. Media Activism and the Internet. Routledge, London (2002)

[Menary, 2007] Menary, R.: Writing as thinking. Language Sciences 29(5), 621–632 (2007)

[Merton, 1965] Merton, R.K.: The Matthew Effect in science: the reward and communications systems of science are considered. Science 159, 56–63 (1965)

[Merton, 1996] Merton, R.K.: The Matthew Effect, ii. In: Sztompka, P. (ed.) On Social Structure and Science, pp. 318–336. Chicago University Press, Chicago (1996)

[Mesoudi et al., 2006] Mesoudi, A., Whiten, A., Dunbar, R.: A bias for social information in human cultural transmission. The British Journal of Psychology 97, 405–423 (2006)

[Metcalf, 2005] Metcalf, R.: Rethinking the ad hominem: A case study of chomsky. Argumentation 19(1), 29–52 (2005)

[Michod and Aurora, 2003] Michod, R.E., Aurora, M.N.: On the reorganization of fitness during evolutionary transitions in individuality. Integrative and Comparative Biology 43, 64–73 (2003)

[Michod and Herron, 2006] Michod, R.E., Herron, M.D.: Cooperation and conflict during evolutionary transitions in individuality. Journal of Evolutionary Biology 19, 1406–1409 (2006)

[Michod, 2005] Michod, R.E.: On the transfer of fitness from the cell to the multicellular organism. Biology and Philosophy 20, 967–987 (2005)

[Michod, 2007] Michod, R.E.: Evolution of individuality during the transition from unicellular to multicellular life. Proceedings of the National Academy of Sciences, USA 104, 8613–8618 (2007)

[Milgram, 1974] Milgram, S.: Obedience to Authority. An Experimental View. Harpercollins, New York (1974)

[Milner and Goodale, 1995] Milner, A.D., Goodale, M.A.: The Visual Brain in Action. Oxford University Press, Oxford (1995)

[Milner et al., 2001] Milner, A.D., Dijkerman, H.C., Pisella, L., McIntosh, R.D., Tilikete, C., Vighetto, A., Rossetti, Y.: Grasping the past: delay can improve visuomotor performance. Current Biology 11, 1896–1901 (2001)

[Mintzberg, 1989] Mintzberg, H.: A note on that dirty word 'efficiency'. In: Mintzberg, H. (ed.) On Management. Inside Our Strange World of Organizations, pp. 330–334. The Free Press, New York (1989)

[Mitchell and Beach, 1990] Mitchell, T.R., Beach, L.R.: 'Do i count thee? Let me count'. Toward an understanding of intuitive and automatic decision making. Organizational Behavior and Human Decision Processes 47, 1–20 (1990)

[Mithen, 1996] Mithen, S.: The Prehistory of the Mind. A Search for the Origins of Art, Religion and Science. Thames and Hudson, London (1996)

[Mithen, 1999] Mithen, S.: Handaxes and ice age carvings: hard evidence for the evolution of consciousness. In: Hameroff, A.R., Kaszniak, A.W., Chalmers, D.J. (eds.) Toward a Science of Consciousness III. The Third Tucson Discussions and Debates, pp. 281–296. MIT Press, Cambridge (1999)

[Moll et al., 2002] Moll, J., Oliveira-Souza, R., Eslinger, P.J., Bramati, I.E., Mourao-Miranda, J., Andreiuolo, P.A., Pessoa, L.: The neural correlates of moral sensitivity: A functional magnetic resonance imaging investigation of basic and moral emotions. The Journal of Neuroscience 22(7), 2730–2736 (2002)

[Moody, 2001] Moody, G.: Rebel Code. Linux and the Open Source Revolution. Perseus Publishing, New York (2001)

[Moor and Bynum, 2002] Moor, J.H., Bynum, T.W. (eds.): Cyberphilosophy. Blackwell, Malden, MA (2002)

[Natsoulas, 2004] Natsoulas, T.: To see is to perceive what they afford: James J. Gibon's concept of affordance. Mind and Behaviour 2(4), 323–348 (2004)

[Nelson and G.Winter, 1972] Nelson, R.R., Winter, S.G.: An Evolutionary Theory of Economic Change. Belknap Press, Cambridge, Mass. (1972)

[Neumann and Morgenstern, 1944] von Neumann, J., Morgenstern, O.: The Theory of Games and Economic Behavior. Princeton University Press, Princeton (1944)

[Newell and Simon, 1972] Newell, A., Simon, H.A.: Human problem solving. Prentice-Hall, Englewood Cliffs, NJ (1972)

[Norman, 1988] Norman, D.: The Design of Everyday Things. Addison Wesley, New York (1988)

[Norman, 1993] Norman, D.: Things that Make us Smart. Addison Wesley, New York (1993)

[Norman, 1999a] Norman, D.: Affordance, conventions and design. Interactions 6(3), 38–43 (1999)

[Norman, 1999b] Norman, D.A.: The Invisible Computer. The MIT Press, Cambridge, MA (1999)

[Norman, 2002] Norman, J.: Two visual systems and two theories of perception: an attempt to reconcile the constructivist and ecological approaches. Behavioral and Brain Sciences 25, 73–144 (2002)

[Nussbaum, 2001] Nussbaum, M.C.: Upheavals of Thought. The Intelligence of Emotion. Cambridge University Press, Cambridge (2001)

[Oatley, 1992] Oatley, K. (ed.): Best Laid Schemes: the Psychology of Emotions. Cambridge University Press, Cambridge (1992)

[Odling-Smee et al., 2003] Odling-Smee, F.J., Laland, K., Feldman, M.W.: Niche Construction. A Neglected Process in Evolution. Princeton University Press, New York, NJ (2003)

[Odling-Smee, 1988] Odling-Smee, J.J.: The Role of Behavior in Evolution. Cambridge University Press, Cambridge (1988)

[Okasha, 2006] Okasha, S.: Evolution and the Levels of Selection. Oxford University Press, Oxford (2006)

[Osofsky et al., 2005] Osofsky, M.J., Bandura, A., Zimbardo, P.G.: The role of moral disengagement in the execution process. Law and Human Behavior 29(4), 193–209 (2005)

[Pata, 2009] Pata, K.: Revising the framework of knowledge ecologies: how activity patterns define learning spaces? In: Lambropoulos, N., Romero, M. (eds.) Educational Social Software for Context-Aware Learning: Collaborative Methods & Human Interaction. Information Science Reference, pp. 241–267. Hershey, New York (2009)

[Patokorpi, 2008] Patokorpi, E.: Simon's paradox: Bounded rationality and the computer metaphor of the mind. Human Systems Management 27, 285–294 (2008)

[Peirce, 1931–1958] Peirce, C.S.: Collected Papers of Charles Sanders Peirce. In: Collected Papers of Charles Sanders Peirce, Harvard University Press, Cambridge, MA (1931/1958); Hartshorne, C., Weiss, P. (eds.) vol. 1-6; Burks, A.W. (ed.) vol. 7-8

[Peirce, 1967] Peirce, C.S.: The Charles S. Peirce Papers: Manuscript Collection in the Houghton Library. The University of Massachusetts Press, Worcester, MA (1967), Annotated Catalogue of the Papers of Charles S. Peirce. Numbered according to Richard S. Robin. Available in the Peirce Microfilm edition. Pagination: CSP = Peirce / ISP = Institute for Studies in Pragmaticism

[Pepper, 2000] Pepper, J.W.: Relatedness in group-structured models of social evolution. Journal of Theoretical Biology 206, 355–368 (2000)

[Pepper, 2007] Pepper, J.W.: Simple models of assortment through environmental feedback. Artificial Life 13(1), 1–9 (2007)

[Perry, 2003] Perry, M.: Distributed cognition. In: Carroll, J. (ed.) HCI, Models, Theories, and Frameworks. Morgan Kaufman, London (2003)

[Pinker, 2003] Pinker, S.: Language as an adaptation to the cognitive niche. In: Christiansen, M.H., Kirby, S. (eds.) Language Evolution, pp. 16–40. Oxford University Press, Oxford (2003)

[Pisella et al., 2006] Pisella, L., Binkofski, F., Lasek, K., Toni, I., Rossetti, Y.: No double-dissociation between optic ataxia and visual agnosia: multiple sub-streams for multiple visuo-manual integration. Neuropsychologia 44(13), 2734–2748 (2006)

[Polanyi, 1966] Polanyi, M.: The Tacit Dimension. Routledge & Kegan Paul, London (1966)

[Popper, 1945] Popper, K.R.: Open Society and its Enemies. Routledge, London (1945)

[Putnam, 2000] Putnam, R.: Bowling Alone. Simon & Schuster, New York (2000)

[Raab and Gigerenzer, 2005] Raab, M., Gigerenzer, G.: Intelligence as smart heuristics. In: Sternberg, R.J., Prets, J.E. (eds.) Cognition and Intelligence. Identifying the Mechanisms of the Mind, pp. 188–207. Cambridge University Press, Cambridge, MA (2005)

[Rader and Vaughn, 2000] Rader, N., Vaughn, L.: Infant reaching to a hidden affordance: evidence for intentionality. Infant Behavior and Development 23, 531–541 (2000)

[Raftopoulos, 2001a] Raftopoulos, A.: Is perception informationally encapsulated? The issue of theory-ladenness of perception. Cognitive Science 25, 423–451 (2001)

[Raftopoulos, 2001b] Raftopoulos, A.: Reentrant pathways and the theory-ladenness of perception. Philosophy of Science 68, S187–S189 (2001); Proceedings of PSA 2000 Biennal Meeting (2000)

[Ramachandran and Hirstein, 1997] Ramachandran, V.S., Hirstein, W.: Three laws of qualia: what neurology tells us about the biological functions of consciousness. Journal of Consciousness Studies 4, 429–457 (1997)

[Raymond, 2001] Raymond, E.S.: The Cathedral and the Bazar. O'Reilly, Sebastopol, CA (2001)

[Raymond, 2004] Raymond, E.S.: Open minds, open source. Analog 1(8), 23–30 (2004)

[Reed, 1988] Reed, E.S.: James J. Gibson and the Psychology of Perception. Yale University Press, New Haven, CT (1988)

[Remagnino et al., 2005] Remagnino, P., Foresti, G., Ellis, T.: Ambient Intelligence. A Novel Paradigm. Springer, Berlin (2005)

[Rest, 1986] Rest, J.: Moral Development: Theory. Praeger, New York (1986)

[Richerson and Boyd, 1998] Richerson, P.J., Boyd, R.: The evolution of human ultrasociality. In: Eibl-Eibesfeldt, I., Salter, F.K. (eds.) Indoctrinability, Ideology, and Warfare, Evolutionary Perspectives, pp. 71–95. Berghahn Books, New York (1998)

[Richerson and Boyd, 2005] Richerson, P.J., Boyd, R.: Not by Genes Alone. How Culture Trasformed Human Evolution. The University of Chicago Press, Chicago and London (2005)

[Rock, 1982] Rock, I.: Inference in perception. In: PSA. Proceedings of the Biennial Meeting of the Philosophy of Science Association, vol. 2, pp. 525–540 (1982)

[Rose, 2005] Rose, S.: The Future of the Brain. The Promise and Perils of Tomorrow's Neuroscience. Oxford University Press, Oxford (2005)

[Rupert, 2010] Rupert, R.D.: Cognitive Systems and the Extended Mind. Oxford University Press, Oxford/New York (2010)

[Russ-Eft, 2004] Russ-Eft, D.: Ethics in a global world: an oxymoron? Evaluation and Program Planning 27, 349–356 (2004)

[Sabini and Silver, 1982] Sabini, J., Silver, M.: Moralities of Everyday Life. Oxford University Press, Oxford (1982)

[Salomon, 1993] Salomon, G. (ed.): Distributed Cognitions: Psychological and Educational Considerations. Cambridge University Press, Cambridge (1993)

[Scarantino, 2003] Scarantino, A.: Affordances explained. Philosophy of Science 70, 949–961 (2003)

[Schnider, 2001] Schnider, A.: Spontaneous confabulation, reality monitoring, and the limbic system: A review. Brain Research Reviews 36, 150–160 (2001)

[Schultz *et al.*, 2003] Schultz, R.T., Grelotti, D.J., Klin, A., Kleinman, J., Van der Gaag, C., Marois, R., Skudlarski, P.: The role of the fusiform face area in social cognition: Implications for the pathobiology of autism. Philosophical Transactions of the Royal Society, Series B 358, 415–427 (2003)

[Secchi, 2010] Secchi, D.: Extendable Rationality. Understanding Decision Making in Organizations. Springer, New York (2010) (in press)

[Sellen and Harper, 2002] Sellen, A.J., Harper, R.H.R.: The Myth of the Paperless Office. The MIT Press, Cambridge, MA (2002)

[Selten, 1998] Selten, R.: Features of experimentally observed bounded rationality. European Economic Review 42, 413–436 (1998)

[Shakun, 2001] Shakun, M.F.: Unbounded rationality. Group Decision and Negotiation 10, 97–118 (2001)

[Shelley, 1996] Shelley, C.: Visual abductive reasoning in archaeology. Philosophy of Science 63(2), 278–301 (1996)

[Simon, 1947] Simon, H.: Administrative Behavior. Free Press, New York (1947)

[Simon, 1955] Simon, H.A.: A behavioral model of rational choice. The Quarterly Journal of Economics 69, 99–118 (1955)

[Simon, 1959] Simon, H.A.: Theories of decision-making in economics and behavioral science. The American Economic Review 49(3), 253–283 (1959)

[Simon, 1977] Simon, H.A.: Models of Discovery and Other Topics in the Methods of Science. Reidel, Dordrecht (1977)

[Simon, 1978] Simon, H.A.: Rationality as process and a product of thought. American Economic Review 68, 1–14 (1978)

[Simon, 1979] Simon, H.A.: Rational decision making in business organizations. American Economic Review 69, 493–513 (1979)

[Simon, 1983] Simon, H.: Reason in Human Affairs. Stanford University Press, Stanford (1983)

[Simon, 1990] Simon, H.A.: A mechanism for social selection and successful altruism. Science 250(4988), 1665–1668 (1990)

[Simon, 1993] Simon, H.A.: Altruism and economics. The American Economic Review 83(2), 156–161 (1993)

[Simon, 2002] Simon, L.D.: Democracy and the Internet. Allies or Adversaries? Woodrow, Washington, DC (2002)

[Simon, 2005] Simon, H.A.: The Sciences of the Artificial. The MIT Press, Cambridge, MA (2005)

[Singer, 1972] Singer, P.: Famine, affluence, and morality. Philosophy and Public Affairs 1(3), 229–244 (1972)

[Sinha, 2006] Sinha, C.: Epigenetics, semiotics, and the mysteries of the organism. Biological Theory 1(2), 112–115 (2006)

[Small, 2004] Small, H.: On the shoulders of Robert Merton: Towards a normative theory of citation. Scientometrics 60(1), 71–79 (2004)

[Smith, 2004] Smith, A.: The Theory of the Moral Sentiments. Penguin, London (2004) (Originally published in 1759)

[Sperber and Mercier, 2010] Sperber, D., Mercier, H.: Reasoning as a social competence. In: Landemore, H., Elster, J. (eds.) Collective Wisdom. Cambridge University Press, Cambridge (2010) (in press)

[Spohn, 1996] Spohn, W.C.: Who counts? Images shape our moral community. Issues in Ethics 7(2), 229–244 (1996)

[Stearns, 2007] Stearns, S.C.: Are we stalled part way through a major evolutionary transition from individual to group? Evolution 61(10), 2275–2280 (2007)

[Sterelny, 2005] Sterelny, K.: Made by each other: organism and their environment. Biology and Philosophy 20, 21–36 (2005)

[Stich, 2007] Stich, S.: Evolution, altruism and cognitive architecture: A critique of Sober and Wilson's argument for psychological altruism. Biology and Philosophy 22(2), 267–281 (2007)

[Stigler, 1961] Stigler, G.J.: The economics of information. Journal of Political Economy 69, 213–225 (1961)

[Stoffregen, 2003] Stoffregen, T.A.: Affordances as properties of the animal-environment system. Ecological Psychology 15(3), 115–134 (2003)

[Strevens, 2006] Strevens, M.: The role of the matthew effect in science. Studies in History and Philosophy of Science 37, 159–170 (2006)

[Sunstein, 2005a] Sunstein, C.: Conformity and dissent, Public Law and Legal Theory Working Paper No. 34, University of Chicago, Chicago (2005)

[Sunstein, 2005b] Sunstein, C.: Infotopia. How Many Minds Produce Knowledge. Oxford University Press, Oxford (2005)

[Sunstein, 2007] Sunstein, C.: Republic.com 2.0. Princeton University Press, Princeton (2007)

[Susi and Ziemke, 2001] Susi, T., Ziemke, T.: Social cognition, artefacts, and stigmergy. Journal of Cognitive Systems Research 2, 273–290 (2001)

[Sutcliffe, 2003] Sutcliffe, A.: Symbiosis and synergy? Scenarios, task analysis and reuse of HCI knowledge. Interacting with Computers 15, 245–263 (2003)

[Sutherland, 2000] Sutherland, N.S.: Irrationality. The Enemy Within. Constable, London (2000)

[Szatkowska et al., 2007] Szatkowska, I., Szymajska, O., Bojarski, P., Grabowska, A.: Cognitive inhibition in patients with medial orbitofrontal damage. Experimental Brain Research 181(1), 109–115 (2007)

[Thagard, 1988] Thagard, P.: Computational Philosophy of Science. The MIT Press, Cambridge, MA (1988)

[Thagard, 2000] Thagard, P.: Coherence in Thought and Action. The MIT Press, Cambridge, MA (2000)

[Thagard, 2005] Thagard, P.: Mind: Introduction to Cognitive Science, 2nd edn. The MIT Press, Cambridge, MA (2005)

[Thagard, 2010] Thagard, P.: How brains make mental models. In: Magnani, L., Carnielli, W., Pizzi, C. (eds.) Model-Based Reasoning in Science and Technology. Abduction, Logic and Computational Discovery. Springer, Berlin (2010)

[Thomas, 1999] Thomas, N.J.T.: Are theories of imagery theories of imagination? An active perception approach to conscious mental content. Cognitive Science 23, 207–245 (1999)

[Tindale, 2007] Tindale, C.W.: Fallacies and Argument Appraisal. Cambridge University Press, Cambridge (2007)

[Todd and Gigerenzer, 2003] Todd, P.M., Gigerenzer, G.: Bounding rationality to the world. Journal of Economic Psychology 24, 143–165 (2003)

[Tooby and DeVore, 1987] Tooby, J., DeVore, I.: The reconstruction of hominid behavioral evolution through strategic modeling. In: Kinzey, W.G. (ed.) Primate Models of Hominid Behavior, pp. 183–237. Suny Press, Albany (1987)

[Turvey and Shaw, 2001] Turvey, M.T., Shaw, R.E.: Toward an ecological physics and a physical psychology. In: Solso, R.L., Massaro, D.W. (eds.) The Science of the Mind: 2001 and Beyond, pp. 144–169. Oxford University Press, Oxford (2001)

[Tversky and Kahneman, 1983] Tversky, A., Kahneman, D.: Extensional versus intuitive reasoning: The conjunction fallacy in probability judgment. Psychological Review 90, 293–315 (1983)

[Tylén, 2007] Tylén, K.: When agents become expressive: a theory of semiotic agency. Cognitive Semiotics 1, 84–101 (2007)

[Van der Veer Martens and Goodrum, 2006] Van der Veer Martens, B., Goodrum, A.A.: The diffusion of theories: A functional approach. Journal of the American Society for Information Science and Technology 57(3), 330–341 (2006)

[Verbeek, 2008] Verbeek, P.-P.: Cyborg intentionality: Rethinking the phenomenology of human technology relations. Phenomenology and the Cognitive Sciences 7, 387–395 (2008)

[Verbeek, 2009] Verbeek, P.-P.: The moral relevance of technological artifacts. In: Sollie, P., Duwell, M.D. (eds.) Evaluating New Technologies, pp. 63–77. Springer, Berlin (2009)

[Vicente, 2003] Vicente, K.J.: Beyond the lens model and direct perception: toward a broader ecological psychology. Ecological Psychology 15(3), 241–267 (2003)

[Vrij, 2008] Vrij, A.: Detecting Lies and Deceit Pitfalls and Opportunities. Wiley, New York (2008)

[Walton, 1980] Walton, D.: Why is the ad populum a fallacy? Philosophy and Rhetoric 13, 264–278 (1980)

[Walton, 1995] Walton, D.N.: Arguments from Ignorance. Penn State University Press, Philadelphia (1995)

[Walton, 1997] Walton, D.: Appeal to Expert Opinion: Arguments from Authority. Penn State Press, University Park (1997)

[Walton, 1999a] Walton, D.: Ethotic arguments and fallacies: The credibility function in multi-agent dialogue systems. Pragmatics & Cognition 7(1), 177–203 (1999)

[Walton, 1999b] Walton, D.: Rethinking the fallacy of hasty generalization. Argumentation 13(2), 161–182 (1999)

[Walton, 19998] Walton, D.: Ad Hominem Arguments. The University of Alabama Press, Tuscaloosa (1998)

[Walton, 2004] Walton, D.: Abductive Reasoning. The University of Alabama Press, Tuscaloosa, AL (2004)

[Wark, 2004] Wark, M.: A Hacker Manifesto. Harvard University Press, Cambridge, MA (2004)

[Warren, 1995] Warren, W.H.: Constructing an econiche. In: Flach, J., Hancock, P., Caird, J., Vicente, K.J. (eds.) Global Perspective on the Ecology of Human-Machine Systems, pp. 210–237. Lawrence Erlbaum Associates, Hillsdale, NJ (1995)

[Watley and May, 2004] Watley, L.D., May, D.R.: Enhancing moral intensity: The roles of personal and consequential information in ethical decision-making. Journal of Business Ethics 50, 105–126 (2004)

[Weber, 2004] Weber, S.: The Success of Open Source. Harvard University Press, Cambridge, MA (2004)

[Wells, 2002] Wells, A.J.: Gibson's affordances and Turing's theory of computation. Ecological Psychology 14(3), 141–180 (2002)

[Wheeler, 2004] Wheeler, M.: Is language and ultimate artifact? Language Sciences 26, 693–715 (2004)

[Whiten and Byrne, 1997] Whiten, A., Byrne, R.: Machiavellian Intelligence II: Evaluations and Extensions. Cambridge University Press, Cambridge (1997)

[Williams, 2002] Williams, S.: Free as in Freedom. O'Reilly, Sebastopol, CA (2002)

[Wilson and Sober, 1994] Wilson, D.S., Sober, E.: Reintroducing group selection to the human behavioral sciences. Behavioral and Brain Sciences 17(4), 585–654 (1994)

[Wilson and Wilson, 2007] Wilson, D.S., Wilson, E.O.: Rethinking the theoretical foundation of sociobiology. Quarterly Review of Biology 82(4), 327–348 (2007)

[Wilson et al., 2000] Wilson, D.S., Wilczynski, C., Wells, A., Weiser, L.: Gossip and other aspects of language as group-level adaptations. In: Heyes, C., Huber, L. (eds.) Cognition and Evolution, pp. 347–365. The MIT Press, Cambridge, MA (2000)

[Wilson et al., 2002] Wilson, D.S., Wilczynski, C., Wells, A., Weiser, L.: Gossip and other aspects of language as group-level adptations. In: Hayes, C., Huber, L. (eds.) The Evolution of Cognition, pp. 347–365. The MIT Press, Cambridge, MA (2002)

[Wilson, 103] Wilson, R.A.: Wide computationalism. Mind 411, 351–372, 103

[Wilson, 1977] Wilson, D.S.: Structured demes and the evolution of group-advantageous traits. The American Naturalist 111, 157–185 (1977)

[Wilson, 2001] Wilson, R.: Group-level cognition. Philosophy of Science 68(3), S262–S273 (2001)

[Wilson, 2002a] Wilson, D.S.: Evolution, morality and human potential. In: Scher, S.J., Rauscher, F. (eds.) Evolutionary Psychology. Alternative Approaches, pp. 55–70. Kluwer Academic Publishers, Dordrecht (2002)

[Wilson, 2002b] Wilson, D.S.: Darwin's Cathedral. Evolution, Religion, and the Nature of Society. Chicago University Press, Chicago (2002)

[Wilson, 2004] Wilson, R.A.: Boundaries of the Mind. Cambridge University Press, Cambridge (2004)

[Wilson, 2005] Wilson, R.A.: Collective memory, group minds, and the extended mind thesis. Cognitive Processing 6, 227–236 (2005)

[Wilson, 2006] Wilson, D.S.: Human groups as adaptive units: toward a permanent consensus. In: Laurence, S., Carruthers, P., Stich, S. (eds.) The Innate Mind: Culture and Cognition. Oxford University Press, Oxford (2006)

[Windsor, 2004] Windsor, W.L.: An ecological approach to semiotics. Journal for the Theory of Social Behavior 34(2), 179–198 (2004)

[Woods, 2004] Woods, J.: The Death of Argument. Kluwer Academic Publishers, Dordrecht (2004)

[Woods, 2005] Woods, J.: Epistemic bubbles. In: Artemov, S., Barringer, H., Garcez, A., Lamb, L., Woods, J. (eds.) We Will Show Them: Essays in Honour of Dov Gabbay, vol. II, pp. 731–774. College Publications, London (2005)

[Woods, 2009] Woods, J.: Seductions and Shortcuts: Error in the Cognitive Economy (2009) (forthcoming)

[Yerkovich, 1977] Yerkovich, S.: Gossiping as a way of speaking. Journal of Communication 27, 192–196 (1977)

[Young, 2006] Young, G.: Are different affordances subserved by different neural pathways? Brain and Cognition 62, 134–142 (2006)

[Zerubavel, 2006] Zerubavel, E.: The Elephant in the Room. Silence and Denial in Everyday Life. Oxford University Press, Oxford (2006)

[Zhang and Patel, 2006] Zhang, J., Patel, V.L.: Distributed cognition, representation, and affordance. Cognition & Pragmatics 14(2), 333–341 (2006)

[Zhang, 1997] Zhang, J.: The nature of external representations in problem solving. Cognitive Science 21(2), 179–217 (1997)

[Zukow-Goldring and Arbib, 2007] Zukow-Goldring, P., Arbib, M.A.: Affordances, effectivities, and assisted imitation: Caregivers and the directing of attention. Neurocomputing 70(13-15), 2181–2193 (2007)

Index

Cognitive Systems Monographs

Edited by R. Dillmann, Y. Nakamura, S. Schaal and D. Vernon